Taming an Uncertain Future

Future Perfect: Images of the Time to Come in Philosophy, Politics and Cultural Studies

Series editors: Michael Marder, IKERBASQUE Research Professor of Philosophy, University of the Basque Country, Spain, and Patricia Vieira, Associate Professor, Spanish and Portuguese, Georgetown University, USA.

The *Future Perfect* series stands at the intersection of critical historiography, philosophy, political science, heterodox economic theory, and environmental thought, as well as utopian and cultural studies. It encourages an interdisciplinary reassessment of the idea of futurity that not only holds a promising interpretative potential but may also serve as an effective tool for practical interventions in the fields of human activity that affect entire countries, regions, and the planet as a whole.

Titles in the Series

The Future of Europe: Democracy, Legitimacy and Justice After the Euro Crisis edited by Serge Champeau, Carlos Closa, Daniel Innerarity and Miguel Poiares Maduro

Taming an Uncertain Future: Temporality, Sovereignty, and the Politics of Anticipatory Governance by Liam Stockdale

The Politics of Virtue: Post-liberalism and the Human Future by John Milbank and Adrian Pabst

Taming an Uncertain Future

Temporality, Sovereignty, and the Politics of Anticipatory Governance

Liam P. D. Stockdale

ROWMAN &
LITTLEFIELD
INTERNATIONAL

London • New York

Published by Rowman & Littlefield International Ltd
Unit A, Whitacre Mews, 26-34 Stannary Street, London SE11 4AB
www.rowmaninternational.com

Rowman & Littlefield International Ltd. is an affiliate of Rowman & Littlefield
4501 Forbes Boulevard, Suite 200, Lanham, Maryland 20706, USA
With additional offices in Boulder, New York, Toronto (Canada), and Plymouth (UK)
www.rowman.com

British Library Cataloguing in Publication Data
A catalogue record for this book is available from the British Library

ISBN: HB 978-1-78348-500-0
PB 978-1-78348-501-7

Library of Congress Cataloging-in-Publication Data

Names: Stockdale, Liam P. D.
Title: Taming an uncertain future : temporality, sovereignty, and the politics of anticipatory governance / Liam P. D. Stockdale.
Description: Lanham : Rowman & Littlefield International, [2016] | Series: Future perfect | Includes bibliographical references and index.
Identifiers: LCCN 2015037058| ISBN 9781783485000 (cloth : alk. paper) | ISBN 9781783485017 (pbk. : alk. paper) | ISBN 9781783485024 (electronic)
Subjects: LCSH: Political planning—Philosophy. | Time—Political aspects. | Exceptionalism.
Classification: LCC JF1525.P6 S73 2016 | DDC 320.601—dc23 LC record available at http://lccn.loc.gov/2015037058

∞ ™ The paper used in this publication meets the minimum requirements of American National Standard for Information Sciences—Permanence of Paper for Printed Library Materials, ANSI/NISO Z39.48-1992.

Printed in the United States of America

Contents

Acknowledgments

Any undertaking so simultaneously arduous and rewarding as the publication of one's first book cannot be completed alone. I therefore owe significant debts of gratitude to many people who have encouraged and assisted me in this process over the past several years.

I will begin by thanking the editors of the *Future Perfect* book series, Patricia Vieira and Michael Marder, whose positive response to my inquiries about this exciting series were crucial to my decision to publish here. They have been supportive of this project throughout its development, for which I am grateful. I am similarly thankful to Sarah Campbell and Sinéad Murphy at Rowman & Littlefield, whose consistent editorial assistance and support was crucial to helping me navigate this first foray into the book publishing process. Patricia Stevenson, also of Rowman & Littlefield, has been extremely helpful on the production side, and I am also grateful to Robert and Cynthia Swanson for indexing services.

I first began to develop the ideas explored in this book during my time as a doctoral student in the political science department at McMaster University in my hometown of Hamilton, Ontario. Among my mentors at that outstanding institution, I am especially thankful to Peter Nyers and Marshall Beier, whose constructive engagements with earlier versions of this book's arguments have significantly shaped the manuscript's final form. But it is to Tony Porter—whose support for this project at all stages and attendant investment in my own development as a junior scholar has been invaluable—that I owe even greater thanks. Without Tony's mentorship, I would certainly not be in the fortunate position of composing these words. I also owe a great debt to a number of former graduate school colleagues and current great friends. Mark Busser, Mike Di Gregorio, Philippe Frowd, Heather Johnson, Jessica Merolli, Jen Mustapha-Vanderkooy, Nicole Wegner, Mark Williams, and Katie Winstanley have all provided me with the vital combination of insightful professional engagement and warm personal friendship, without which I would not have made it this far. I must also thank Barbara Falk from the Canadian Forces College, without whose initial encouragement I might never have pursued the doctoral work on which this book is based, and whose generous "outsider" advice and ongoing friendship have been of great help in my navigation of the occasionally treacherous waters of academia.

This manuscript was completed over the course of postdoctoral fellowships at McMaster's Institute on Globalization and the Human Condition (IGHC) and the McMaster Institute for Innovation and Excellence in Teaching and Learning (MIIETL), respectively. I am grateful for the support of both institutions, and particularly that of Don Goellnicht and Lihua Qian at IGHC and Zafar Syed and Leeanne Romane at MIIETL, whose encouragement and support have been crucial in creating a positive working environment where this project could be brought to fruition. I am also grateful for the financial support of the Social Sciences and Humanities Research Council of Canada—in the form of a Joseph Armand-Bombardier Canada Graduate Scholarship—and the McMaster Faculty of Social Sciences—in the form of a Dr. Harry Lyman Hooker Senior Fellowship—both of which have funded elements of the research contained in this book.

But it is to my family that I owe the greatest thanks. My mother and father, Claudia and Rick Stockdale, passed along their own expansive intellectual curiosity and passion for learning and education, and they have supported me unreservedly as I followed where my own interests took me. My sister, Arielle Stockdale, has inspired me with her meaningful work outside the academy while always being there with a kind and encouraging word as I continue to toil within it. Finally, my wonderful spouse and partner Jenn—who has endured every plodding step of this undertaking alongside me—deserves more gratitude than it is possible to convey here. This book is dedicated to all four of you, since its completion would simply not have been possible without your ongoing kindness, understanding, love, and support.

Liam Stockdale
Hamilton, Ontario
November 2015

Chapter One

Introduction

Temporality, Futurity, and the Political

In one of the more memorable passages from his *Confessions*, Saint Augustine muses perplexedly about the nature of time. "What, then, is time?" he asks, before proceeding to offer what is perhaps the pithiest articulation of the paradoxical relationship between human subjectivity and temporality found in the canon of Western philosophy: "if no one asks me, I know; if I wish to explain it to one that asketh, I know not" (1968, 40). This at once simple and profound observation captures well the point that while all human subjects are in some way fundamentally aware of time, it remains perpetually beyond our capacity to fully grasp, and thus eludes cogent conceptual articulation. Indeed, even a cursory parsing of the voluminous philosophical literature on the subject reveals that, on the one hand, time is recognized as central to the most basic questions of intellectual inquiry and human existence—"[it] is a fundamental aspect of all that occurs, a boundary condition on phenomena" (Turetzky 1998, xi). On the other hand, it also becomes apparent that there likely exist as many temporal understandings as there are philosophical orientations; as many theoretical articulations of time as there are theorists to articulate them (see, for instance, Adam 2004; Bender and Wellbery 1991; Elias 1992; Gale 1968; Grosz 1999; Koselleck 1985; McClure 2005; McCumber 2011; Turetzky 1998). Thus, although humanity is unlikely to develop a universally accepted understanding or conceptualization of time, the point remains that our relations to time form a fundamental part of the human condition. As Kimberly Hutchings, paraphrasing Kant, puts it, "Time . . . conditions all our experience of ourselves" (2008, 3).

Yet our interactions with time are by no means wholly harmonious. Quite the contrary, as it is our status as beings in time that is a primary source of the difficulties and insecurities that in many ways define the human experience. As Bonnie Honig argues, "Time and man . . . are agonistically related," in that the free activities of human subjects "interrupt would-be time sequences," while the vicissitudes of time in turn impinge upon human freedom (Honig 2008, 108). In the context of this agonic relationship between humanity and temporality, the desire to manage the latter by exerting some degree of influence over it emerges as a critical imperative in human affairs. Thus, what social theorist Barbara Adam refers to as a "quest for time control" can be detected to varying extents across all forms of social and cultural production—from philosophic, to religious, to political activity—such that the social world as we know it is constituted to a significant extent by concepts and practices that are intimately concerned with the management, governance, and even mastery of time itself (Adam 2004, 19–20, 124, 152; see also Luhmann 1982, 274). Time is thus fundamental to shaping the human experience, and the imperative to control, or at least govern, time can be understood as a crucial undercurrent in the ongoing constitution and operation of the social world.

These two insights might initially appear to relate primarily to broad philosophical questions about the character of existence and the nature of being; however, they also serve as a point of departure for this book—a study whose subject matter is not the phenomenological or eschatological but the political. Indeed, in the chapters that follow, I am interested in critically exploring how temporality more broadly—and the imperative to control the unfolding of the future in particular—is embedded in the epistemic foundations and practical operations of contemporary rationalities and mechanisms of societal governance.

A TEMPORAL LENS

The analytical importance of time has long been recognized across the disciplines of the social sciences and humanities, whose primary subject matter is the realm(s) of human affairs. Indeed, prominent works from the past several decades in anthropology (such as Johannes Fabian's *Time and the Other* [1983]); sociology and social theory (such as Barbara Adam's *Time and Social Theory* [1994]); history and historiography (such as Hayden White's *Metahistory* [1973]); and (continental) philosophy (such as Paul Ricoeur's *Time and Narrative* [1990])—among many others—have all dealt explicitly with humanity's relationship to time in their respective disciplinary contexts. In the study of politics more specifically, theorists have long recognized the centrality of temporal questions to the elemental issues of societal governance with which they are concerned. Major works of political theory such as John Gunnell's *Political Philosophy and Time* (1987) and Sheldon Wolin's *Politics and Vision* (2004) directly address the relationship between time and how we think about the political. To be sure, interest in time has ebbed and flowed with the intellectual currents of disciplinary inquiry, but an underlying sensitivity to the importance of taking time seriously has nonetheless characterized the scholarly ethos of these areas of study whose subject matter is the world created by humanity.

Moreover, we are currently witnessing a resurgent interest in questions of time more broadly, and futurity more specifically, among scholars in these and related fields. Most often prompted by normative concerns relating to such issues as the perceived acceleration of life in late modernity (Hassan and Purser 2007; Rosa and Scheuerman 2009), the specter of ecological collapse wrought by the increasingly unsustainable lifestyles many now take for granted (Atwood 2008; Bastian 2012), or the emergence of other ostensibly new threats whose catastrophic potentialities seem to destabilize our traditional relationships to the future (Beck 2002, 2008), a wide variety of authors are beginning to train their gazes upon the temporalities of the social world by adopting what can be termed a "temporal lens" that "puts time . . . front and centre" in their analyses thereof (Ancona et al. 2001, 645). This scholarly trend is particularly constructive for examinations of contemporary politics, in that some of the most significant political dynamics of the current moment stem from shifts in the broader political imagination which have brought explicitly temporal questions very much to the fore (see Agathangelou and Killian 2016; Lundborg 2012; Hutchings 2008). For instance, the growing influence of narratives proclaiming the onset of a "world risk society"—characterized by the erosion of our ability to control the unfolding of the future to an extent that affords us an adequate degree of ontological certainty in the present, and exemplified by such inherently global problems as climate change, financial crises, and transnational terrorism—suggests that time in general, and the irruptive contingency of the future in particular, have become discursively framed as pressing problems that must be actively addressed through political channels (Beck 1999). This discourse is mirrored in practice by the widespread emergence of societal governance strategies that are explicitly oriented toward taming an uncertain future through anticipatory action in the present (Anderson 2010a; de Goede and Randalls 2011). Such strategies—which are often described in terms of "risk management," "prevention," "preemption," and so forth—can thus be understood as "attempts to control time," and are becoming increasingly prevalent in myriad areas of human affairs across the globe (Kessler 2011, 2181). This combination of an epistemic shift toward a focus on overtly temporal problems and a concurrent practical shift toward future-oriented governmental logics has thus in many ways "reconfigured the politics of space into a politics of time," such that the temporal has (re-)emerged as a primary site and subject of governance (Kessler 2011, 2181).

Of course, the novelty of this emergent "temporalization" of the political should not be overstated, as the political realm has been concerned with controlling the unfolding of time and taming the contingency of the future since long before Machiavelli exhorted his prince to master *fortuna*. Yet this ongoing (re-)framing of temporal contingency as a political problem—and the attendant proliferation of strategies and mechanisms of government developed more explicitly for this purpose—suggests the need for in-depth engagement with the politics of time and futurity in general, and logics of anticipatory governance in particular. A growing number of scholars across multiple disciplines have begun to critically interrogate certain specific elements of these wider issues; however, there is still much work to be done with respect to theorizing the *idea* of anticipatory governance in primarily abstract, conceptual terms—particularly regarding its attendant implications for the organization and exercise of political power in the liberal democratic polities that are at the forefront of its enactment. A primary contribution of this study, therefore, is to provide a book-length engagement with precisely such questions.

The remaining chapters thus develop a sustained exploration and critique of the logics underlying the anticipatory political rationalities whose global proliferation has made it increasingly important to foreground questions of temporality and futurity in political analysis. The term "political rationality" is used here to denote a "discursive field within which the exercise of power is conceptualised," which combines "justifications for particular ways of exercising power" with "notions of the appropriate forms, objects, and limits of politics" (Rose and Miller 1992, 175). A political rationality can thus be understood as the epistemic framework that guides political action in a certain context by both articulating the sort(s) of problem(s) to be addressed and providing a program for action through which political power can be mobilized toward these ends. This book seeks to explore and explain the sort of politics that anticipatory rationalities aimed at taming an increasingly uncertain future make possible when deployed as mechanisms of societal governance. In this respect, the core argument hinges on a claim that the practical implementation of anticipatory political rationalities requires a paradigm of political power closely reminiscent of that which is associated with a politics of "exceptionalism," as theorized most notably by Carl Schmitt, Giorgio Agamben, and their subsequent interlocutors. Political exceptionalism describes a condition in which juridical limitations on the actions of a sovereign authority are diminished to the point of practical irrelevance. In other words, it signifies a political circumstance characterized by "serious distortions in the restraining effects that the rule of law . . . [has] on the arbitrary exercise of power," such that those with the capacity to deploy sovereign power are effectively placed beyond the law (Huysmans 2004, 327). This book develops the argument that such a form of political authority is enacted by anticipatory rationalities of governance because the latter are concerned with controlling time by acting upon *potential* futures, and since these futures are inherently unknowable, a highly arbitrary form of political decision-making that relies upon imagination and speculation is required to make acting upon them practically possible. The chapters of part II in particular emphasize that the paradigm of political power required by such preemptive logics of action conspicuously resembles that which underlies a politics of exceptionalism, thus suggesting that mechanisms of anticipatory governance have the potential to significantly alter the character of political subjectivity, particularly when implemented in a liberal democratic polity. In short, therefore, this book foregrounds temporality and futurity by developing a critical theorization of the exceptional forms of political power that are enacted by the logics of anticipatory governance whose global proliferation makes such a foregrounding of temporality and futurity important in the first place.

An in-depth engagement with the politics of anticipatory governance requires extensive analysis of particular instances in which these sorts of strategies and mechanisms are enacted in practice. While any attempt to explore all the areas of human affairs in which they have become prevalent could not be accomplished in a single volume, it is nonetheless possible to develop conceptual insights relevant to a broader critique of anticipatory governance through a more detailed examination of one particularly prominent manifestation thereof. This is the path followed by this book, as the subsequent chapters develop a thorough engagement with perhaps the most conspicuous example of anticipatory governance in the contemporary global political context: the widespread deployment of "preemptive" strategies toward the governance of (in)security. Indeed, the post-9/11 rise of transnational terrorism as the dominant issue in the global security imagination has placed temporal questions more generally, and the taming of the future's contingency more specifically, at the core of the contemporary politics of (in)security (Anderson 2010b). This is because the specter of terrorism has become largely framed as a radically irruptive, catastrophic potentiality inhabiting an ultimately unknowable security future that can thus only be adequately governed through anticipatory interventions aimed at stopping the proverbial "next attack" before it occurs. The proliferation of such future-oriented rationalities thus constitutes perhaps the most notable development in the post-9/11 global security climate (Dershowitz 2006; Ericson 2008), and also represents an archetypical example of the future-oriented "temporalization" of the political discussed above. Accordingly, this book's wider concern with highlighting the importance of foregrounding temporality and futurity in political analysis and more specifically exploring what is at stake with the rise of anticipatory governance rationalities will be pursued through an in-depth critique of how the problem of temporal contingency has been prioritized within the global security imagination and responded to through the development of security strategies premised upon governing the future through preemptive interventions in the present.

To put all of this another way, the project developed in this book is oriented around three principal tasks, which can be understood as follows: (1) emphasizing that questions of temporality and futurity should be given greater attention in the study of (global) politics, and providing a book-length illustration of how this might be done which especially highlights the critical potential of doing so; (2) establishing a foundation for a thorough conceptual critique of anticipatory governance rationalities, whose global proliferation provides a primary analytical impetus for taking temporality and futurity more seriously; and (3) offering a comprehensive critical interrogation of one particular manifestation of such rationalities in practice—namely, the preemptive regime(s) of (in)security governance that have emerged within the context of the post-9/11 global War on Terror. That all three tasks are principally concerned with conceptual questions is indicative of the primarily theoretical character of the book, as the subsequent chapters collectively represent an attempt to critically think through how questions of temporality and futurity—and the imperative to exert agentic control over both—are crucially embedded in contemporary modes of societal governance. It should also be noted that these three tasks do not directly correspond to the book's chapter divisions, but rather are interwoven throughout the various arguments developed therein.

Framing this book in the preceding way is useful because it helps to highlight its key contributions. In this respect, the concern with demonstrating how we might take temporality and futurity more seriously in political analysis contributes to broader disciplinary debates in political science and international relations (IR) regarding the proper subject and scope of scholarship. Indeed, this book seeks not only to show what adopting a "temporal lens" for the study of the political might look like but also to emphasize that doing so is both methodologically prudent—since questions of temporality and futurity increasingly underpin the key dynamics of contemporary (global) politics—and analytically productive—since doing so enables the development of innovative readings of key conceptual and practical issues. This hints at the contributions embodied by the second and third tasks mentioned above, which relate to the critical potential of foregrounding temporality and futurity in this way. In this regard, the book's attempt to think through the implications of the ongoing "temporalization" of the political through the rise of anticipatory governance mechanisms elicits significant critical insights into the way political power is organized and exercised in the contemporary moment. Specifically, by tracing how the resurgent imperative to govern time through future-oriented political rationalities requires what amounts to a politics of exceptionalism in order to be functionally implemented, the arguments developed in this book provide the epistemic basis for a thorough critique of anticipatory governance as an idea on both conceptual and normative grounds. Moreover, because these arguments are developed through a sustained engagement with the realm of global (in)security governance, the book also contributes to the critical literature in IR and security studies in two notable ways. First, by suggesting that exceptionalist politics are presupposed by the logic of anticipatory governance itself, it offers an innovative, conceptually oriented intervention that contributes to explanations for the widespread proliferation of exceptionalist practices in the post-9/11 era—a topic that has been of paramount interest to critical security scholars over the past decade (Neal 2009). Secondly, by arriving at these insights through an in-depth theorization of how anticipatory rationalities have been applied to the governance of (in)security in particular, this study develops a highly comprehensive conceptualization of what has been termed "preemptive security" in the context of contemporary global politics (de Goede 2008, 162; Stockdale 2013; Sullivan and Hayes 2010). My understanding and use of this term follows the existing literature on the subject, in which "preemptive security" refers generally to "security practices that aim to act on threats that are unknown and recognized to be unknowable, yet deemed potentially catastrophic, requiring security intervention at the earliest possible stage" (de Goede, Simon, and Hoijtnik 2014, 412). While much critical security scholarship has been concerned with problematizing the various state practices and policies that fall under this classification (Amoore 2014a), the extant literature has devoted less attention to describing and unpacking this approach to (in)security governance in more general or abstract conceptual terms. Filling this gap thus represents a key contribution of this book.

The book's aims should thus be broadly understood in the spirit of self-identified "critical" scholarship, which seeks to "re-open assumptions that have grounded our political thought" and develop alternative understandings of pressing political issues and concepts in the contemporary context (Edkins and Vaughan-Williams 2009, 2). However, it is particularly important to recognize that the critical insights developed in the following chapters are in many ways ultimately the result of the analysis being conducted through an explicitly temporal lens. In this respect, this book shows that a move to take temporality and futurity seriously is demanded by the emergent dynamics of the current (global) political moment, and doing so will also productively facilitate criticality in the study thereof.

A NOTE ON METHOD AND APPROACH

As is perhaps already clear, the chief contributions this book seeks to make are almost entirely conceptual in nature. Indeed, the subsequent chapters consist primarily of in-depth exercises in theoretical argumentation and conceptual analysis that are the result of sustained critical reflection upon the issues in question. With respect to questions of method, therefore, this book deliberately draws upon the precedent set by Hedley Bull when he prefaced *The Anarchical Society*—which would become one of the most important works in the history of IR scholarship—by describing it as "an attempt to deal with a large and complex subject simply by thinking it through" (Bull 1977, xiii). While I have no pretensions about this book achieving a similar level of renown to Bull's, my effort to grapple with how questions of temporality and futurity, and logics of anticipatory governance embedded in the constitution and operation of contemporary (global) politics should be understood as employing his "method" in this respect.

Yet, while principally concerned with conceptual problems and arguments, there is also an empirical component to the analysis, as I make references of varying depth throughout the text to practical manifestations of anticipatory governance strategies in the context of contemporary global security. It should be stressed, however, that these direct engagements with such instances are undertaken not for the purpose of developing any detailed empirical account thereof, but rather to further flesh out conceptual arguments relating to how anticipatory governance mechanisms—of which these practices are prototypical examples—operate. In other words, the purpose of the empirical components is to provide a degree of concrete grounding for the conceptual claims that comprise the core of the book's contribution, and should thus be understood as a series of illustrative examples that have been chosen on the basis of their explanatory capacity. It is worth briefly identifying some of the more prominent examples with which I engage to better demonstrate how they are utilized in this way.

- Chapter 3's discussion of the qualitatively unique character of post-9/11 applications of preemption to questions of (inter)national security is partially illustrated through the Obama administration's use of "predictive assessments" in the creation of flight "watchlists" that ban individuals from travelling by air in the United States. Subsequently, what I describe as the inherently productive aspects of this contemporary politics of preemptive security are illustrated in part through an examination of the American FBI's counterterror strategy which uses paid informants to recruit individuals suspected of terrorist sympathies for the purposes of goading them into devising plots that can then be preemptively foiled.

- Part of chapter 4's conceptual analysis of the logic of preemption as a political rationality considers how it compresses the temporal window of political decision-making. This point is illustrated with reference to both the character of American political debate around the passage of the USA PATRIOT Act in the immediate aftermath of the 9/11 attacks and the 2005 killing of Jean Charles de Menezes in London's Stockwell underground station. The latter example is also used to illustrate the role of affect as a driver of anticipatory decision-making, particularly in the security context.
- Chapter 5's examination of the relationship between the logic of preemption and the politics of exceptionalism utilizes several illustrative examples to flesh out core conceptual claims. In this respect, the chapter opens with an in-depth parsing of US Attorney General Eric Holder's important 2012 speech aimed at justifying the Obama administration's targeted killing policy. The subsequent discussion of the epistemic centrality of imagination to a politics of preemption is fleshed out through an examination of Somali-Canadian Mohamed Hersi's preemptive arrest and ultimate conviction in Canada for the offense of "attempting to participate in a terrorist activity." The subsequent discussion of "counter law" as an illustration of how preemptive security suspends the juridical order considers several examples, including the implementation of UN Security Council Resolution 1373, the European Union's financial "blacklisting" regime used to target suspected channels of terrorist financing, and the anti-democratic character of certain provisions of India's core counterterrorist legislation. Finally, the discussion of how preemptive security enacts a "decisionist" paradigm of sovereign authority examines the George W. Bush administration's promotion of the "unitary executive" doctrine—typified by Bush's unprecedentedly frequent use of presidential "signing statements" and subsequently continued in important ways by the Obama administration—and the indefinite detention regime that served as a primary element of the immediate American response to the 9/11 attacks.

- Chapter 6 includes the most in-depth examination of one particular case, as it uses the 2011 targeted killing via drone strike of Anwar al-Awlaki—an American citizen and al-Qaeda militant—to illustrate the chapter's conceptual arguments about the precarious subjectivities enacted under a preemptive security regime.

These are the most prominent illustrative examples employed in the text, though additional references to other events and practices can also be found in the subsequent chapters. And again, while certainly not insignificant to the overarching arguments, these components should be understood as secondary to the conceptual claims and insights that constitute the book's primary contribution.

Time, Space, and Space/Time

Finally, at this point, readers may be skeptical as to whether my intended focus upon futurity and the temporal must come at the expense of a sensitivity to the spatial. Some may go even further and contend that it is illegitimate to speak of space and time as discrete categories, in that the two are so "inextricably interwoven" that we must "insist on . . . the necessity of thinking in terms of space-time" instead (Massey 1994, 261, 269). Such a line of thinking might thus imply that any attempt to analytically prioritize the temporal in the manner attempted by this book is potentially problematic, on both methodological and theoretical grounds.

While I recognize that such arguments are compelling and appreciate the importance of maintaining an adequate sensitivity to the spatial in any analysis of the social world, I do not believe such concerns necessarily undermine the legitimacy of the book's temporally focused analytical approach. Indeed, despite issuing the above invocation, critical geographer Doreen Massey also makes room for such an approach, saying of space and time that "it is not that we cannot make any distinction at all between them, but that the distinction we do make needs to hold the two in tension" (1994, 261). In other words, it is possible to accept the fundamental theoretical point about the inexorable interconnectivity of space and time—and the productive, mutually constitutive tensions between the two in the context of the social world (Crang 2007, 62)—while still prioritizing one over the other for analytical purposes. This should be understood as the approach of this book with respect to the above concerns.

In other words, while an underlying aim is to further emphasize the importance of foregrounding questions of temporality and futurity, in no way should this be construed as an eschewal of the importance of the spatial to the questions and problems that will be examined in the following chapters. The subsequent analysis thus certainly recognizes that the issues considered—which range from the abstract conceptualization of sovereignty to the practical politics of drone warfare—are also intimately bound up with questions of space, and it is not my intention to deliberately play down the significance thereof. In this sense, I am not interested in making the negative case *against* prioritizing the spatial in the study of contemporary societal governance and/or the post-9/11 global governance of (in)security; rather, I am interested in adding to the positive case *for* more prominently foregrounding the temporal more broadly and futurity more specifically.

STRUCTURE OF THE BOOK

The remainder of the book is divided into two parts. Part I, which includes chapters 2 and 3, considers broader theoretical questions about the relationship between temporality, futurity, and the political, particularly pertaining to the proliferation of anticipatory governance more broadly and preemptive approaches to problems of (in)security more specifically. Part II, which includes chapters 4–6, develops an in-depth conceptual critique of the latter by exploring how the logic of preemption has been deployed in the context of the post-9/11 War on Terror. It is worth outlining each chapter in further detail before proceeding.

Chapter 2 opens the analysis by considering the politics of temporal governance from a broader perspective, and thus begins by reading the fundamental political concept of sovereignty through a "temporal lens." Particular emphasis is placed on highlighting the logic of temporal control that is crucial to sovereignty's coherence as a normative political principle, and that has served as a dominant "solution" to the "problem" of contingency in the modern political imaginary. I then consider how this solution is becoming increasingly challenged by the shifting dynamics of the contemporary human experience. Here the discussion takes up the point already mentioned in this introduction that time more generally, and the imperative to govern the future in particular, is occupying an increasingly prominent place in the contemporary sociopolitical imagination, with one result being the proliferation of anticipatory modes of societal governance. After a detailed examination of the similar but importantly distinct concepts of "risk" and "precaution" which underpin the conceptual architecture of such governance rationalities, I consider how the practice of sovereign politics is shifting to reflect a more active and direct approach to temporal control rooted in these and related ideas. The chapter concludes by asserting that this rise of anticipatory governance rationalities has significant implications for the organization and exercise of political power in the contemporary global political context, particularly in the case of the liberal democratic societies that are in the vanguard of this trend. This latter point is the focus of the remaining chapters, which explore what is at stake in this respect by developing an in-depth theorization of the "preemptive security" strategies that characterize the post-9/11 global War on Terror.

This process begins in chapter 3, which examines how the logic of anticipatory governance is manifested in the contemporary global security realm by elaborating on what the idea of "preemptive security" can be understood to mean in the post-9/11 context. Answering this question requires that the idea of preemption be situated historically and conceptually in relation to the realm of (inter)national security. To do so, the discussion contrasts how preemption has been articulated and conceptualized in the more traditional discursive context of international law with the way it has been articulated and manifested in post-9/11 counterterrorism discourses. Through this comparison, it is argued that the contemporary conceptualization of preemption is qualitatively different from previous understandings, primarily in the way it frames radical *un*certainty about the future as the basis for, rather than an impediment to, anticipatory action. The remainder of the chapter then establishes the analytical parameters for the rest of the book by outlining an epistemic framework through which to think critically about the political implications of this formulation of preemption. The question of ontology is addressed first, where it is contended that the post-9/11 idea of preemptive security can be best understood as the sort of "political rationality" discussed earlier in this introduction. The discussion then considers how the relationship between agency and structure operates within the context of such preemptive political rationalities, before moving on to outline the particular understanding of "sovereignty" that is required to adequately comprehend the way political power operates in this context. It is argued that Judith Butler's (2006) notion of "sovereignty within governmentality" and the associated idea of the "petty sovereign" are of particular importance to both of these considerations, and these ideas are discussed in detail. The chapter then concludes part I by speaking to the scope of the analysis undertaken in the remainder of the book, specifically addressing issues related to potential charges of Eurocentrism vis-à-vis the sort of general conceptual analysis to which it aspires. Having thus laid the foundation for an in-depth theorization of preemptive security as an exemplar of anticipatory governance, this task is undertaken in part II.

Chapter 4 opens part II by developing a detailed examination of how preemption operates as a political rationality, using its mobilization in the contemporary security context as an exemplar. The central argument here returns more overtly to the theme of temporality by contending that a politics of preemption functions by manipulating our experiences of time itself. In particular, it is argued that by constructing all future possibilities as potentially imminent and thus subject to anticipatory action, the logic of preemption fundamentally rearticulates our subjective relation to the future. In other words, a preemptive politics "makes the future present" by granting the merely potential, virtual future a significant degree of causal purchase in the decision-making of the actual, lived present. These points are then further unpacked by exploring their implications for questions of concrete political praxis. Here it is argued that the logic of preemption compresses the timescape of political decision-making by demanding immediate action to avoid an always potentially imminent catastrophe. This process of temporal compression in turn prioritizes the affective responses, or "gut feelings," of designated decision-makers over broader public deliberation, thus enacting antidemocratic modes of governance. This suggests that there are further significant implications for the organization and exercise of political power under a regime of anticipatory governance.

Chapter 5 unpacks this idea in much greater detail by exploring how preemptive approaches to (in)security governance in particular enact a form of sovereign authority that conspicuously resembles a paradigm of political "exceptionalism." The argument once more hinges on questions of temporality and futurity, as the key point is that a political rationality premised upon anticipatorily governing the future must rely upon an epistemic foundation of speculative knowledge created through the exercise of the imagination rather than the analysis of empirically verifiable fact. This has the effect of vesting within the designated authority a radically enhanced degree of discretionary subjectivity, since the necessity of dealing in speculation about potentialities that may never come to pass precludes the straightforward application of mechanisms of juridical oversight and democratic negotiation to a politics of preemption. Such an emancipation of decision-making from the circumscriptions of legal and political norms—and the attendant creation of a "decisionist" paradigm of sovereignty—are also defining characteristics of political "exceptionalism," thus suggesting an originary connection between anticipatory governance and exceptionalist politics. As described above, these conceptual claims are fleshed out in considerable depth through a number of illustrative examples taken from the post-9/11 prosecution of the global War on Terror.

Chapter 6 then considers the implications of the originary link between preemption and exceptionalism for the experience of political subjectivity, as it is argued that a preemptive security regime brings into being a political condition characterized by a pervasively precarious subjective condition. This is because the exceptionalist paradigm of sovereign authority that it presupposes effectively eliminates any juridical mediation between sovereign and subject, thus rendering the latter perpetually vulnerable to what amounts to arbitrary interventions by the former. This argument is developed through a discussion of the Obama administration's drone warfare program, with a particular emphasis upon the 2011 targeted killing of al-Qaeda operative and US citizen Anwar al-Awlaki. This is a particularly instructive case, since it represents precisely the sort of exceptional practice whose continual possibility is a necessary condition for enacting preemptive approaches to (in)security governance. The discussion then takes a step back to consider how the arguments developed thus far undermine the conceptual coherence of preemption as a political/security rationality. Here it is argued that the precarious subjectivities enacted by the exceptional politics of preemption closely resemble the sort of condition that it is normatively premised upon diminishing. I then conclude by arguing that these constitutive tensions can be productively understood through the lens of Jacques Derrida's concept of "autoimmunity."

The seventh and concluding chapter brings the book to a close by considering how the preceding analysis of preemptive security in particular suggests that the proliferation and normalization of an exceptionalist politics and its attendant precarious subjectivities constitutes precisely what is at stake with the rise of anticipatory governance more broadly. It is argued that this poses potentially serious ethicopolitical problems for the liberal democracies that are implementing such strategies across a variety of policy areas beyond the security realm, since the anticipatory logic upon which they are based appears manifestly incompatible with certain fundamental principles of a democratic society. The modifications to the organization and exercise of political power that are presupposed by the constitutive logics of anticipatory governance must therefore be given serious consideration in public policy debate relating to the implementation of such strategies, with a proper weighing of these implications against the powerful imperative to tame an increasingly uncertain future.

Part One

Time is a sort of river of passing events, and strong is its current; no sooner is a thing brought to sight than it is swept by and another takes its place, and this too will be swept away.

—Marcus Aurelius, *Meditations*, Book IV

Make use of time, let not advantage slip.

—Shakespeare, *Venus and Adonis*

Always in motion is the future.

—Yoda, *The Empire Strikes Back*

Chapter Two

The Politics of Temporal Control

As discussed in the introduction, our relation to time conditions all aspects of the human experience to at least some degree—human being is being in time. While neither this chapter nor this book are explicitly concerned with the philosophical treatment of time as such, the centrality of time and attempts to control it to the conceptual architecture of the social world provides something of a foundational point for the subsequent analysis. Indeed, as Niklas Luhmann suggests, questions of time are fundamentally integral to the processes by which the social world is both created by and rendered intelligible to human subjects—time is a crucial "aspect of the social construction of reality" (Luhmann 1982, 274; Opitz and Tellmann 2015, 109). It thus follows that concepts central to that constructive process might be (re-)read to emphasize their temporal inflections in such a way that opens up critical space for productively rethinking their role in the organization of human affairs. This chapter begins by following such a trajectory, with a particular focus on a concept that is crucial to the ordering of contemporary society: sovereignty.[1] In this respect, the first section employs a self-consciously "temporal lens"[2] to develop a conceptual reading of sovereignty that highlights how its constitutive normative promises are premised upon a logic of temporal control that has provided a powerful "solution" to the "problem" of temporal contingency in the governance of human affairs. While certainly not a novel claim, this is an important point to consider as a departure point for exploring the politics of anticipatory governance, since the apparent challenges posed to this solution by the emergent exigencies of late modern societies have been crucial to the widespread proliferation of explicitly future-oriented political rationalities. The remainder of the chapter considers this point in more detail by exploring how shifts in the contemporary political imagination have led to the (re-)emergence of temporal contingency as a significant political problem, with a key consequence being the development of a more overtly "temporalized" politics characterized by the integration of the logic of risk and its more radical offshoot precaution into mechanisms of societal governance. This discussion thus lays the foundation for the remainder of the book, which critically interrogates the implications of these developments for the organization and exercise of political power, particularly in the context of the liberal democratic polities at the forefront of these trends.

SOVEREIGNTY AND THE GOVERNANCE OF TIME

The concept of sovereignty as a principle of political organization is most easily understood in terms of space. Indeed, the spatial demarcation of a parcel of territory over which a particular figure or entity wields ultimate political authority is the conceptual core of the idea of sovereignty (Kurtulus 2005, vii), while the spatialized image of a clearly delineated border definitively separating a politically controllable "inside" from a contingently anarchic "outside" comprises its most paradigmatic concrete representation (Walker 1993). Moreover, the spatiality of sovereignty has been crucial to its entrenchment as a conceptual foundation for modern thinking about the political in that, from the birth of Western political thought at the Athenian *polis*, the very idea of politics has been bound up with the attendant need for a space in which its practice can properly take place. As Scott Nelson puts it, "Western political theory has consistently developed political ideals on the basis of political conceptions pertaining to a well-bounded space where an accepted, unquestioned practice of politics was thought to rightly locate itself" (2010, 7). When considered as a spatial concept, therefore, sovereignty provides a compelling response to this fundamental topological necessity of politics.

However, as many notable contemporary theorists of sovereignty have emphasized (see Walker 1991, 1993, 2001), an understanding of the concept developed primarily through a spatial lens is inadequate, since sovereignty also embodies a logic that is inherently temporal in nature. A key consideration in this respect is the rather basic point that the impetus for spatially demarcating a sovereign boundary—which stems from its ostensible capacity to create a secured space within which a progressive, potentially emancipatory politics becomes possible—also implies that this space be maintained and protected *through time* since, to put it bluntly, "politics takes time" (Manning 2004, 65; Wolin 1997). In other words, the more overt, normatively grounded spatial aspects of sovereignty rely upon more implicit imperatives relating to the governance of time. The following discussion explores this idea in more detail by reading the concept of sovereignty in a manner that highlights the importance of a particular logic of temporal control to its conceptual constitution and coherence as a political principle.

It is useful to begin this process by considering what might be understood as the normative foundations of sovereignty. Indeed, by addressing elemental questions of social organization, sovereignty is a fundamentally normative concept that does not so much describe an extant condition as prescribe an ongoing course of action. Notably writing "in defense of sovereignty," W. J. Stankiewicz affirms that the concept consists of "premises [and] assertions about the direction that man chooses (rather than is compelled) to follow . . . about man's beliefs about what 'ought' to be, what is possible, and the steps necessary to realize his beliefs" (1969, 31). It is these premises and assertions that constitute the normative foundations of sovereignty—since they are what have enabled the idea of sovereignty to become attractive as a political principle—and they are rooted in the *promises* that theories of sovereignty make regarding the capacity to bring into being the best possible world through the arrangement of human social relations along the prescribed lines.

In this respect, at the core of sovereignty's normative foundation is a claim to offer a practical framework for the establishment and maintenance of a stable socio-political order (Stankiewicz 1969, 4, 10, 14–15). As Stankiewicz emphasizes, "[W]hat cannot be ignored is the function of sovereignty in maintaining social order: it is this function which creates the obedience that makes sovereign power and its exercise possible" (1969, 10). Put slightly differently, most theorists of sovereignty recognize it as a social technology concerned with taming the forces of contingency and anarchy that are alleged to characterize human existence through the vesting of ultimate political authority in a supreme locus of power within a bounded space (Hoffman 1998, 4; Schmitt 2005; Stankiewicz 1969, 15). This can be seen most simply, for instance, in the use of various state-of-nature thought experiments in classic contractarian theories of sovereignty, or in Carl Schmitt's reference to the unpredictability of what he calls "real life" when outlining his authoritarian theory of sovereign decisionism (Schmitt 2005, 14–15). International relations theorist James Der Derian articulates the normative basis of sovereignty pithily: "It falls upon the sovereign state to protect us from the turbulence of nature and anarchy that permanently lies in wait offshore and over the horizon" (2008, 282). A central assumption is that by cordoning off a rigidly bounded enclave from a dangerous outside, sovereignty makes possible the pursuit of the good life on the inside by creating and maintaining a climate of secure, controlled stability (Jackson 2007, 114; Walker 2010, 66; 1993, 42). The normative foundation of sovereignty is thus premised on a claim that it is only within the protected bastion of certainty and regularity created by the inscription of a sovereign boundary that "aspirations to the good, the true, and the beautiful may be realizable" (Walker 1993, 62). The coherence of sovereignty as a political principle is in this sense rooted in a promise to "enabl[e] citizens to realize their humanity within" the securely bounded space that it provides (Walker 2010, 253), and the attraction of this promise underlies the sovereign state's persistence as "the basic authority assumption, the underlying premise, of modern politics and law" (Jackson 2007, 11; Nelson 2010, 18; Onuf 1991). The logic of sovereignty promises to guarantee "something elementary" for the successful operation of human society by instituting "stable, familiar, and convenient modalities of social expectation, interaction, and co-operation" (Jackson 2007, 160), and thereby providing "the solution to all troubles, theological, ontological, political; or at least as the primary solution that would in turn permit other solutions through government and law" (Walker 2010, 192). In short, sovereignty constitutes a means to slay the malicious demons of anarchical contingency and establish a climate of stable certainty that enables the creation and maintenance of a space within which the ends of human progress can be pursued through the proper practice of politics.

However, it is important to note that this explicit emphasis on order and stability which is central to the normative foundations of sovereignty also contains implicit assumptions concerning those forces and exigencies that conceptions of sovereignty are formulated *in response to*, or as bulwarks *against*. Such considerations are crucial to understanding how a logic of temporal control underpins the broader normative foundations of sovereignty, and thus merit further consideration here. In this regard, a key point is that for the purported "benefits" of sovereignty—namely, the maintenance of sociopolitical order through the creation of an enduringly stable, knowable, controllable space—to be understood as normatively valuable and worth pursuing through the establishment of a sovereign order, their antitheses (such as anarchy, contingency, uncertainty, instability, and the like) must be assumed to be inherently pernicious and thus worth taking concerted measures against (Bartelson 1995, 101). While such assumptions often go unmentioned in particular theories of sovereignty, the inherent dangers of a world characterized by these elements either are overtly emphasized or can be clearly read into the intellectual contexts of particular works—as in the seminal treatments by Bodin or Hobbes that emerged in the midst of pervasive sociopolitical turmoil wrought by the breakdown of established systems of order. Whether implicit or explicit, however, such a valuation of one set of social conditions over its converse is a key element of sovereignty's normative foundation, since claims about the necessity of erecting a sovereign boundary to resist the forces of anarchy and contingency logically imply that the infiltration by these forces into the bounded sovereign space at any point would critically endanger the potential for actualizing any eudaimonistic potentials therein. And since this actualizing process necessarily takes time, the sovereign boundary's resistance to these forces must also endure through time for its normative promises to be kept. In other words, to remain conceptually coherent as a normative political principle, the logic of sovereignty must include the capacity to actively guard against its antitheses not only comprehensively but also indefinitely. As Nelson puts it, sovereignty is a response to "the age-old problem of securing peace, order, and stability and insuring the longevity of these commodities as ultimate ends" (Nelson 2010, 32). Indeed, the normative promises of sovereignty would mean very little if its spatial demarcations were not buttressed by the promise of their own temporal endurance, since the practice of politics and the attendant pursuit of any conception of the good life therein must occur over time.

The importance of this capacity to endure through time thus constitutes the most basic aspect of sovereignty's constitutive logic of temporal control. However, more important in this regard is the extent to which the antithetic forces against which the demarcation of the sovereign boundary ostensibly protects are themselves functions of the flow of time, since this emphasizes that sovereignty is underpinned by a logic of temporal control premised more specifically upon the governance of time's inherently irruptive contingency. A brief engagement with Machiavelli can help clarify this point, since he has written perhaps the most famous work theorizing the politics of confronting and managing the vicissitudes of time. To be sure, Machiavelli does not use the term "sovereignty" in his writings, which may lead some to dismiss the use of his thought in any discussion of the concept; however, especially in *The Prince*, his arguments prefigure in important ways elements of what would later be explicitly articulated in theories of sovereignty—particularly regarding the governance of temporal contingency—and are thus useful for helping explain sovereignty's constitutive logic of temporal control.

Most notable in this respect is that Machiavelli's foundational idea of *fortuna*—against which the proto-sovereign prince must continually struggle—encapsulates how the flow of time embodies the specter of uncertainty, contingency, and instability, which are precisely the forces from which subsequent theories of sovereignty promise to provide respite (Machiavelli 1995, 74–77; Pocock 2003, 38). In his noted reading of *The Prince*, for instance, Pocock describes how, for Machiavelli, politics "is the art of dealing with *fortuna* as the force which directs [contingent] events and thus symbolizes pure, uncontrolled, and unlegitimated contingency"; as such, the governing prince's *virtù* stems in significant part from a capacity to bring this contingency under control within the domain of his political authority (Pocock 2003, 156, 178). This is further emphasized in chapter III of *The Prince* itself, where Machiavelli rejects the received wisdom of passively "tak[ing] advantage of the passage of time" in the context of establishing a cohesive and governable polity, asserting instead the importance of following the example of the Romans, who "looked ahead and took action to remedy problems before they developed" (Machiavelli 1995, 11; Pocock 2003, 161, 198–99). He also later contends that "a ruler who depends entirely on his good fortune will be destroyed when his luck changes"—an occurrence that is itself necessarily a function of the passage of time (Machiavelli 1995, 75–76). Importantly, Machiavelli's perspective on these questions is grounded in his broader conceptualization of time as radically contingent and thus in need of governance through "strength [*virtù*] and prudence, for *in time, anything can happen*, and the passage of time brings good mixed with evil and evil mixed with good" (Machiavelli 1995, 11, emphasis added). Pocock also emphasizes this point by speaking to Machiavelli's belief that "the world is unstabilized and the unexpected a constant threat," and thus that *fortuna* "would destroy you if you did not . . . act in time" (2003, 178, 166). For Machiavelli, then, the unfolding of time is characterized by a radical contingency, and the capacity to tame this contingency is a key element of a successful polity as guided by the *virtú*-ous prince. Indeed, only then can the sort of stable social order that, in his famous discussion of the relative virtues of cruelty and compassion, Machiavelli identifies as necessary for fulfilling the potentials of human existence become possible (Machiavelli 1995, 51).

Machiavelli's political temporality thus suggests that "men are not predisposed to accept a world of becoming"—implying an existence plagued by the uncertainties of the open, unknown future—"but are instead more at home in the world of stasis"—implying a condition in which socio-political stability makes possible a thriving existence premised upon more than mere survival (Nelson 2010, 39). A primary struggle of the prince and the polity he embodies is thus against the vicissitudes of time itself, and a failure to adequately confront and potentially master its contingencies will lead to the breakdown of political order and thus ruin for both (Bartelson 1995, 113; Nelson 2010, 22; Walker 1993, 39). However, that such a mastery of time through the *virtú*-ous actions of the prince is carried out within the boundaries of his domain of power is also crucial. Indeed, while the most powerful forces against which the prince's political agency is to be mobilized are embedded in the flow of time, the successful mobilization of this agency necessarily requires a defined area in which the prince's authority is paramount—in other words, a spatially demarcated territorial unit. Thus, while they emerged after his own time, explicit theories of sovereignty express Machiavelli's own perspective, as described by R. B. J. Walker, that "time is understood as a problem to be overcome . . . by fixing a home for man in space—the state" (Walker 1993, 40). This is, in essence, the logic of temporal control that underpins the normative foundations of sovereignty.

The vicissitudes of *fortuna* against which Machiavelli's prince must prevail in order to create an enduringly stable polity are thus closely analogous to the elements of temporal contingency against which later theories of sovereignty claim to insulate human subjects. Both lines of thought imply that this can be most effectively achieved through the vesting of ultimate power in a particular figure within a bounded political community over which this figure exercises primary authority. Machiavelli's theorization of political temporality thus helps to illustrate that sovereignty's normative promises are inherently reliant upon a degree of temporal control manifesting as the capacity to endure through time, and the very forces against which these promises claim to protect are themselves fundamentally embedded in the process of time's unfolding. Indeed, "[s]overeignty is the concept that performs the work of precision amidst considerable ambiguity and flux" (Nelson 2010, 146), and it is based on a core assumption that "temporality can be fixed and tamed within the spatial co-ordinates of territorial jurisdictions" (Walker 1993, 14). It is this taming of temporality through the parcelling off of a space from the recurrently anarchic, contingent, uncertain temporalities of pre- or non-sovereign human existence that ultimately enables the emergence of a controllable, progressive temporality on the inside and thus ostensibly facilitates the realization of the emancipatory political project implied by the normative claims of sovereignty (Bartelson 1995, 101; Hutchings 2008, 30; Nelson 2010, 93; Walker 2010, 66). The sovereign border can therefore be understood not only as a spatial demarcation but also as a temporal boundary, as its inscription promises to provide a means through which time itself can be sufficiently governed and controlled by human subjects.

While perhaps making no novel revelations about the logic of sovereignty as such, focusing on the temporal underpinnings of its normative foundations in this way can help us think productively about broader global political processes specifically related to the changing dynamics of sovereignty. For instance, as a number of prominent social and political theorists have argued at length, the emergent exigencies of late modern societies are fundamentally disrupting many of the established certainties and regularities previously made possible by the organization of human affairs in accordance with such fundamental sociopolitical concepts as sovereignty (Beck 2008, 2005; Virilio 2010). With respect to the latter in particular, these disruptions are rooted in an apparent undermining of sovereignty's logics of control by such spatio-temporally "de-bounded" hazards as financial crises wrought by globally mobile capital, environmental destruction wrought by anthropogenic climate change, the rapid worldwide transmission of infectious disease, and catastrophic irruptions of violence wrought by transnational terrorism (Beck 2002, 41). Indeed, the forces and dynamics at the source of such potential catastrophes operate explicitly beyond the neat topological contours of the logic of sovereignty, transcending and traversing the lines of demarcation that have heretofore acted as apparent bulwarks of certainty and stability against irruptions of contingency. By highlighting the importance of temporal control to sovereignty's normative foundation and thus broader conceptual coherence, the reading of sovereignty developed in this section helps bring into focus how the problems associated with these sorts of emergent hazards stem in particular from their disruption of sovereignty's logic of *temporal* governance, in that they compromise our capacity to tame the contingencies embedded in the unfolding of time through the rigid demarcation of a particular space. The next section expands upon this point by pursuing a more detailed discussion of these trends in contemporary world politics—including, in particular, an exploration of how the concept of risk and associated rationalities of anticipatory governance have arisen in response to an emergent desire to recapture a degree of the temporal controllability and attendant ontological certainly promised by the logic of sovereignty.

POLITICS AT THE END OF CERTAINTY

We have thus far seen how "sovereignty can . . . be used as a tool to manage an uncertain future" (Aalberts and Werner 2011, 2185), and in this sense offers a compelling "solution" to the "problem" of temporal contingency in the organization and governance of human affairs. Again, this solution is found in sovereignty's foundational normative claim that the creation, control, and zealous defense of an explicitly delineated spatial domain enables humanity to effectively govern the irruptive excesses of time's unfolding. Sovereignty as a political principle thus comprises an explicitly spatial mechanism for governing temporal contingency, and this somewhat paradoxical solution to the "problem of time" has profoundly influenced modern understandings of political possibility.

Indeed, as Opitz and Tellmann, following Luhmann, contend, "Modernity is linked to an intense *futurization* of the future," which "implies an unbinding from the past" and "an opening towards what is to come"; however, this "intense futurization that characterizes modernity . . . does not occur without a simultaneous rise of modes of temporal control," of which, as the above discussion has shown, the concept of sovereignty is a prominent exemplar

(2015, 111, emphasis original). The conceptual primacy of sovereignty within the modern political imagination (Onuf 1991) has meant that "modern political thought has *depended* on the claim that temporality can and must be tamed by the spatial certainties of sovereign states" (Walker 1993, 178, emphasis added). In other words, sovereignty's logic of temporal control has constituted a fundamental tenet of modern thinking about the political, which has had especially important implications for the way temporality and futurity are understood and addressed as political problems. Most notably in this regard is that the rise of state sovereignty as the dominant principle of political organization (Spruyt 1996) led the question of temporal governance to become largely coextensive with the question of spatial control (Bauman 2000, 110–118). Indeed, sovereignty's constitutive normative claim that time could be adequately tamed through the demarcation and defense of a sovereign boundary—that "spatial order [was] the condition within which temporality might be controlled" (Walker 2010, 254)—emerged as a dominant way of understanding and approaching problems of temporal contingency within the modern sociopolitical imaginary. This meant that such problems did not need to be directly confronted through explicitly temporally inflected or future-oriented political rationalities, since the logic of sovereignty implied that they could be adequately managed—and the mastering of *fortuna* that Machiavelli associated with the singularly *virtù*-ous actions of his ideal prince could be accomplished—simply by upholding the territorial integrity of the sovereign state. In other words, the preeminence of state sovereignty within the modern political imagination served to fold the question of temporal control into the question of spatial governance, thus diminishing temporal contingency as a distinct political problem via the implication that it could be adequately governed through the spatial logics of sovereignty alone.

To be sure, this account of modern political temporality is rather basic, but the point is simply to emphasize that, as a consequence of sovereignty's undisputed emergence as the dominant ordering principle of modern politics, the logic of temporal control through the governance of space upon which it is premised became entrenched as the prevailing response to the "problem" of time in the political management of human affairs. This point is important in the present context because there is a persuasive argument to be made that sovereignty's mode of governing time is becoming increasingly untenable, and thus that matters of temporal contingency are (re-)emerging as discrete and urgent political problems that must be actively confronted in more direct ways. Indeed, it has been widely argued that in the face of the emerging exigencies of what has been variously termed "late modernity" (Giddens 1991), "postmodernity" (Harvey 1991), "new modernity" (Beck 1992), "liquid modernity" (Bauman 2000), and so on, "the hope that temporality may be tamed within the territorial spaces of sovereign states alone is visibly evaporating" (Walker 1993, 155). While certainly not without its own well-deserved criticisms, it is worth elaborating on this line of thought because it provides a provocative starting point for thinking critically about the changing temporal dynamics of contemporary world politics that have precipitated the proliferation of anticipatory governance strategies.

The theoretical basis for an argument that time and its irruptive contingencies are (re-)emerging as serious political problems is found in the oft-cited claims advanced by a number of prominent social and political theorists that certain emergent trends—such as increasing "social acceleration" due primarily to innovations in information, communication, and transportation technology (Rosa 2013) or a spatiotemporal de-bounding of forces and circulations that challenges conventional managerial and governmental logics (Beck 2002, 41–42)—have drastically diminished the capacity of established institutions and mechanisms of control to provide the degree of ontological certainty that was previously considered both possible and necessary in human society (Beck 2008; Bauman 2000). In other words, there is emerging a widespread perception that "the radical contingency of environments . . . necessitates new ways of governing events and life" (Adey, Anderson, and Graham 2015, 7). According to this view, we have entered an epoch that might be characterized by what Nobel laureate Ilya Prigogine has termed the "end of certainty" (Prigogine 1997). Although speaking about developments in the physical sciences that have cast doubt upon the stable temporal determinism that informed scientific inquiry since at least Newton, Prigogine's turn of phrase is an equally apposite descriptor of an emerging social world where prevailing concepts and strategies devised to confront and tame contingency and uncertainty are being perceived as inadequate in the face of the "global complexity" of our "network society"/"high-speed society"/"world risk society" (Beck 1999; Hassan and Purser 2007; Rosa and Scheuerman 2009; Urry 2003).

An especially important consequence of this putative shift is that the phenomenological experience of time among human subjects is being modified, particularly with respect to the ways we experience the present and relate to the future. In particular, as "old certainties fall away," constructions of the future become infused with a radical *un*certainty, with the broader societal capacity to govern emerging contingencies appearing greatly diminished (Beck 2005, 36). The result is that the lived present is experienced in terms of a growing sense of precarity, since the novel "global risks" lurking in the opacity of the future's increasingly uncertain depths suggest a perpetual imminence of catastrophe (Beck 2008, 9–11; Leccardi 2007, 30; Virilio 2010, 7). Time more generally—and its inherently irruptive capacities in particular—is thus seen to be breaking free of their erstwhile subjugation by the various logics of control that have been established in the development of modern societies (Leccardi 2007, 28). As German social theorist Armin Nassehi puts it, "Modernity promised the capacity to shape and control the world and time . . . [b]ut in late modernity, time itself has come to destroy the potential for any form of social or substantial control" (quoted in Rosa 2009, 102–3).

This line of argument has significant resonance with respect to the concept of sovereignty, which, as we have seen, is among the most central of those social technologies developed to govern temporal contingency. Indeed, the aspects of the current moment highlighted by the theorists cited above seriously challenge sovereignty's constitutive normative claims relating to the provision of a temporally stable bastion of certainty in which the emancipatory possibilities of politics can be realized. Again, this is because the dynamic forces and circulations of a globalized world "thoroughly discombobulate modern notions of place and space that still serve as fundamental *grounding* strategies in late modernity" by operating beyond and across the spatial boundaries that constitute the sovereign state, thus transcending the mode of temporal control ostensibly enacted thereby (Nelson 2010, 38, emphasis original). In other words, the spatial demarcations of the sovereign state alone are becoming incapable of fully insulating human subjects from the potential catastrophes embedded in the accelerative exigencies and irruptive contingencies of the contemporary human experience (Aalberts and Werner 2011, 2193; Beck 2002, 41–42; Nelson 2010, 162). Writing over twenty years ago, Walker presciently described this dynamic and its potential implications, asserting that "if contemporary political life is increasingly characterized by processes of temporal acceleration, then we should expect to experience increasingly disconcerting incongruities between new articulations of power and accounts of political life predicated on the early-modern fiction that temporality can be fixed and tamed within the spatial co-ordinates of territorial jurisdictions" (1993, 14). Thus, if the likes of Beck, Bauman, Virilio, and similar thinkers are taken seriously, the sovereign control of space no longer constitutes a sufficient condition for realizing the degree of certainty and stability associated with the adequate control of temporal contingency. Rather, from such a perspective, the exigencies of late modernity have imbued time itself with a degree of radical irruptivity that cannot be adequately tamed through folding the control of time into the governance of space.

At this point, it should be noted that I am not unproblematically accepting the sweeping, epochal proclamations that inform these sorts of conclusions, nor am I unaware of the myriad deficiencies—both methodological and empirical—that have made such lines of theorizing relatively soft targets of critique (Jarvis 2007; Mythen 2005). The point is merely to emphasize that the arguments presented by such theories *do* nevertheless provide a useful departure point for developing a critical understanding of how the question of time in general—and the contingency of the future in particular—are embedded, understood, and responded to in the contemporary political imagination. This is because if any significant challenge to established mechanisms of temporal governance—such as that which the above narrative suggests is characteristic of our current moment—in fact exists, it should precipitate a corresponding emergence of alternative social logics aimed at more actively confronting the resultant increase in perceived uncertainty. Put in terms specifically related to the concept of sovereignty, a serious challenge to its particular solution to the problem of temporal contingency would presumably lead to a move away from political rationalities that take for granted the assumption that "temporality may be tamed within the territorial spaces of sovereign states alone" (Walker 1993, 155), and toward alternative rationalities premised upon acting more directly on time and the future in a manner more explicitly reminiscent of Machiavelli's *virtù*-ous prince. If we have in fact reached the "end of certainty," and confronting and taming temporal contingency remains as serious a human concern as it has always been (Adam 2004), then we should be witnessing shifts in the political imaginary through which the governance of future *un*certainty becomes identified as a vital imperative and political rationalities and interventions aimed at more actively controlling the unfolding of time itself are prioritized. In what amounts to at least partial validation for the lines of thought discussed above, the character of many emerging strategies and mechanisms of societal governance across myriad areas of human affairs suggests that such an epistemic shift *is* in many ways taking place.

Indeed, even a cursory survey of emerging governance trends across numerous issue areas reveals that political activity is becoming increasingly oriented toward a more direct confrontation of temporal contingency in a manner that radically reformulates, or perhaps even moves beyond, sovereignty's spatialized logic of temporal control. In sectors as diverse as financial regulation (Porter 2009), public health management (Cooper 2006), environmental governance (United Nations 1992), crime prevention (Ericson 2007), urban planning (Coaffee 2009), natural disaster management (United Nations 2013), and national security (Department of Homeland Security 2011), a broad array of governance strategies, technologies, and rationalities are being developed and deployed that frame practical interventions specifically in terms of taming a resurgent temporal contingency and recovering an acceptable degree of certainty. Rosa and Scheuerman's broader observation that the earlier "preponderance of space over time" in modernity is being "inverted and ultimately replaced by the dominance of time" is of particular relevance to the contemporary global political climate, since the erstwhile consensus, in which questions of temporal governance could be effectively folded into questions of sovereign spatial governance, appears to be coming undone in the face of the accelerative exigencies of the current moment (Rosa and Scheuerman 2009, 10).

Consequently, it appears that a new politics of time is in the process of materializing, whereby the more active governance of temporal contingency constitutes a primary imperative, and overtly future-oriented, *anticipatory* rationalities of governance are replacing the spatially oriented, reactive strategies that characterized the political ethos of an earlier modernity (Baumann 2000, 110–18). In other words, because they are increasingly perceived to be inadequate for controlling the potentially catastrophic contingencies ostensibly immanent to the present condition, prevailing governmental concepts and mechanisms—such as the logic of sovereignty—are being increasingly reformulated and supplemented by more temporally oriented rationalities premised explicitly upon governing the future through anticipatory interventions in the present. In this respect, we are witnessing the emergence of "a redefined concept of politics, where the observation of political processes has less to do with [sovereign] states per se, but with the management of uncertainty and contingency" (Kessler 2011, 2165–66). This shift in the political imaginary has served to "reconfigure the politics of space into a politics of time" to such an extent that the legitimacy of political authority is increasingly associated not merely with the assertion and successful exercise of power within a particularly bounded space, but with the demonstrable success of "attempts to control time" itself (Kessler 2011, 2181).

Perhaps the clearest indication of this reconfiguration is found in the conspicuous diffusion of the concept of "risk" across the realm of the political. To be sure, the praxis of societal governance has always been concerned with confronting and managing "risks," understood in the colloquial sense as largely synonymous with hazards or threats (Gardner 2009). However, "risk" has a more specific meaning in the context of contemporary social and political theory, referring to a particular governmental approach to the questions of uncertainty and contingency that is premised upon recovering a degree of temporal control in the face of an increasingly uncertain future. The concept of risk, in other words, denotes both a broader conceptual orientation toward the problem of time (and its associated uncertainties) and a specific rationality for action in the face of an unknowable future. Accordingly, as the exigencies of late modernity are perceived to be destabilizing sovereignty's spatialized solution to these problems, and the need for more active strategies of temporal governance has seemingly become more acute, the notion of risk—along with more radical offshoots such as "precaution" (Ewald 2002)—has been drawn out of its original conceptual home in the actuarial and environmental sciences and injected into governance strategies implemented across a much wider swathe of human affairs (Power 2004). As such, the vernacular of risk constitutes a common thread throughout the diverse areas of political activity in which the governance of temporal contingency has (re-)emerged as a primary imperative, since authorities are responding to this imperative by adopting anticipatory strategies premised upon the active management of the future. Risk is thus "indispensable for understanding our times," and can be considered one of the dominant conceptual orientations to the problem of governing temporal contingency in the contemporary context (Garland 2003, 49). Any inquiry into current relationships between temporality, futurity, and the political—which includes the attendant proliferation of anticipatory logics of governance in particular—requires an understanding of risk and its derivatives as conceptualized and applied in contemporary social and political thought, since these are perhaps the foundational ideas underpinning such emerging rationalities, even if the latter often stray significantly from these conceptual parameters in practice.

Governing the Future through the Present: Risk and Precaution

Samuel Johnson is credited with coining the pithy truism that "the future is purchased by the present." Intuitively obvious as this quip may be, the underlying logic—when taken beyond the superficial insinuation of basic cause and effect—is at the core of the concept of risk as understood in contemporary social and political thought. Indeed, risk is fundamentally premised upon the assumption that human subjects are capable of weathering the vicissitudinous storms of time that erupt as the unknown future transfigures into the lived present, and that properly determined interventions in that present can permit us to exert control over this transfiguration and thus radically diminish the uncertainty with which it invariably confronts us. Peter Bernstein makes this point on the first page of his widely read conceptual history of the idea of risk. Here, he identifies risk as directly concerned with the imperative to tame time and affirms that it is ultimately rooted in the premise that "the future is more than the whim of the gods and that men and women are not passive before nature" (Bernstein 1996, 1). Risk as a sociopolitical concept thus begins with this assumption about humanity's capacity to exert agentic influence over the future's unfolding, and offers a particular epistemic framework through which this ability can be harnessed in practice.

In the most general sense, therefore, risk is best understood as a social technology for grappling with the specter of an uncertain future once considered the purview of the gods and now populated by a vast array of potentially catastrophic contingencies largely of our own making (Beck 2008; Bernstein 1996; Lupton 1999, 7). The idea of risk thus ostensibly renders the contingency of the future both "knowable and actionable" by "mak[ing] the unpredictable predictable" and "turn[ing] uncertainties into possibilities" (Aradau et al. 2008, 150; Beck 1999, 139; Bessant et al. 2003, 13). Accordingly, risk represents a particular orientation toward the problem of time that does not merely aspire to resist the associated specter of uncertainty by spatially insulating human subjects from it, as is the case with the logic of sovereignty; rather, the logic of risk is premised upon a more active and direct confrontation of future uncertainty and temporal contingency through targeted anticipatory action in the present. In a sociopolitical environment beset by the types of radical uncertainties that apparently characterize the contemporary world, risk thus represents a compelling idea upon which to base mechanisms of societal governance, as it provides a specific rationality for taking action in the face of the unknown in a manner that permits the recapture of a degree of ontological certainty through the renewal of human control over the unfolding of time.

These conceptual considerations are perhaps most succinctly captured by Nikolas Rose's description of risk as "a family of ways of thinking and acting, involving calculations about probable futures in the present followed by interventions into the present in order to control that potential future" (Rose 2001, 7). However, in addition to concisely describing the philosophical basis of the concept, his use of the term "calculations" addresses a more practical aspect of risk by hinting at the epistemic mechanisms through which it operates. The key point here is that if the anticipatory interventions deployed under the logic of risk are to be based upon anything other than pure speculation,[3] some sort of informational database is required from which to develop precise calculations about both the likelihood of potential futures and the most effective means of influencing their unfolding in the desired way. As such, the sociopolitical praxis of risk requires the ongoing accumulation of what criminologist Pat O'Malley terms "risk knowledge," which refers to any information deemed to be relevant or necessary for the creation of an informed picture of the future (Ericson and Haggerty 2002; O'Malley 2005, 51). Using such established knowledge as a baseline thus enables reasoned, calculative decisions to be made concerning appropriate interventionary action to be undertaken in the present for the purpose of shaping the future in accordance with a particular set of aims. The concept of risk thus projects a broadly technocratic veneer, as its decisional rationality is premised upon the collection and calculative analysis of relevant data, rather than mere conjecture about what the ultimately unknowable future might hold (Hacking 1990).

However, the putatively neutral, scientific character of risk is belied by the point that the identification of those contingencies that must be acted upon—and thus the determination of precisely what can be understood to constitute relevant risk knowledge—is inevitably a highly politicized process of social construction, which Beck refers to as "staging" (Beck 2008, 10; see also Ewald 1991, 1993). Thus, despite its technocratic façade, the logic of risk remains inherently political, as it is inevitably shot through with the prevailing discursive power relations that shape the construction of all knowledge (Foucault 1980). The politicality of risk has been of particular importance in the contemporary context, since the staging process conventionally relies upon relevant groups of experts whose prestige has been called into question by the ongoing destabilization of the prevailing regimes of certainty with which such epistemic communities are generally associated. Indeed, "the changing relationships of trust between individuals and expert institutions" that has accompanied the broader challenge to established modes of governing contingency has created a rather paradoxical situation where the logic of risk is increasingly cast as the proper response to a newly endemic uncertainty, while the conventional sources of risk knowledge are met with growing skepticism because they are associated with the very regimes of certainty that are becoming increasingly discredited by the apparently radical nature of emergent contingencies (MacNaghten 2005, 136). Dubbed by some risk theorists as the "expert/layman controversy" (Taylor-Gooby and Zinn 2006, 35), this epistemological fissure is a crucial problem at the core of the contemporary proliferation of risk as a mode of governance, and is perhaps best exemplified by the fiercely contentious politics of climate change.

While this aspect of risk politics will not be discussed further, it provides a useful segue into a discussion of the limits of risk as a governance mechanism. It is useful to frame such a discussion by clarifying the conceptual relationship between risk and uncertainty, since risk is constituted in response to uncertainty and reaches its limit when confronted by the most extreme manifestations thereof. In both popular and scholarly discourse, risk and uncertainty are often used almost interchangeably in reference to potential irruptions of harm in the future (see Beck 2008, 17; Best 2008, 359; Knight 1946; Lupton 1999, 7). And although the two terms are intimately related, such a conflation erroneously confuses the nature of the connection between them. This connection can be best expressed by the idea that risk constitutes a sociopolitical tool for tackling and managing uncertainty, which itself constitutes a particular epistemic condition experienced by agents in the social world. In other words, the concept of risk is a product of our ongoing existential confrontation with the uncertainty of the future, and is premised upon rendering the future calculable and therefore governable. Thus, while risk ought not to be conflated with uncertainty, it only becomes an intelligible concept in reference to it—as Garland puts it, "risk always exists in the context of uncertainty" (2003, 52).

Yet the nature of this relationship between the two concepts also hints at the limitations of risk's capacities as a tool of (temporal) governance. Indeed, although risk-based governance strategies are adopted in response to diminishing certainty vis-à-vis the open future, the future's very openness renders it *ultimately* unknowable and thus always imbued with a contingent uncertainty that can *never* be fully tamed, even if we possess an extraordinarily detailed catalogue of risk knowledge. In other words, "our capacity to identify, evaluate, and manage uncertain future events is intrinsically limited, even where there is good data" (Garland 2003, 53), since the ultimate unknowability that defines the future as such always leaves open the possibility that a proverbial "black swan" may appear and thus render even the most meticulously calculated attempts at risk-based governance ineffective (Taleb 2010).

Moreover, each particular circumstance to which a risk-based strategy is applied will be unique in the degree to which it is amenable to governance through the calculative logic of risk. There will inevitably be circumstances and problems beset by an uncertainty whose radicality places them beyond the reach of risk's calculative logic, and thus about whose future, to paraphrase Keynes, we simply *cannot* know (Beck 2008, 18). Such problems embody contingencies whose amenability to governance through present anticipatory interventions is ultimately indeterminate given available knowledge capacities. While traditionally represented by such examples as severe natural disasters, which cannot be predicted and over which we can exert little or no control, one of the key insights of Beck's famed "risk society" thesis is that some of the most threatening and radical uncertainties confronted by contemporary societies are, in fact, explicitly produced by the ongoing development of these societies—such as the potential dangers posed by nuclear waste, genetically modified organisms, climate change, financial deregulation, and transnational terrorism (Beck 1999). In such circumstances, the basic logic of risk as described here is largely inapplicable, and more radical modes of temporal governance must be developed and deployed if a concerted effort to confront uncertainty and tame contingency is to be undertaken (Aradau, Lobo-Guerrero, and Munster, 2008, 151). Indeed, it is quite possible that the available risk knowledge relating to both the potential and the character of future catastrophes of the sort represented by the above examples will be manifestly inadequate for the development of a rationally calculated interventionary program which could reduce the likelihood of their occurrence. Such perceptions have led to the emergence of alternative modes of anticipatory action that go beyond the formulation discussed thus far, and draw instead upon a more radical derivative of the logic of risk most commonly known as "precaution" (Ewald 2002; Sunstein 2007).

As with the concept of risk, the idea of precaution begins with the premise that a lack of certainty about a particular future ought not to preclude preventative action in the face of potential catastrophe (Aradau and Van Munster 2007, 102). However, precaution moves beyond risk by seeking to reclaim the possibility of an actionable decision in the context of an uncertainty so extreme that it cannot be mitigated through an empirical knowledge-based, calculative rationality. Indeed, as François Ewald puts it in a seminal essay on the subject, precaution "does not target all risk situations, but only those marked by two principal features: a context of scientific uncertainty on the one hand, and the possibility of serious and irreversible damage on the other" (2002, 283–84). The logic of precaution thus offers a rationality for decision-making under such conditions, as it purports to bring the radically contingent under the control of future-oriented human agents by making anticipatory decisions possible even in the absence of any significant risk knowledge. Precaution thus "results from an ethic of the necessary decision in a context of uncertainty" and, as such, offers a conceptual framework within which such a decision becomes both practically possible and ethically legitimate (Ewald 2002, 294). Put most simply, then, precaution takes the concept of risk to its logical excess, and it is no coincidence that it has been central to modes of governance developed in response to the radical uncertainties and catastrophic contingencies perceived to characterize contemporary human experiences (Ewald 2002, 283; Taylor-Gooby and Zinn 2006, 46).

While the logic of precaution as a mode of governance has something of a contested genealogy, it is generally agreed to have originated in its contemporary form with the development of the "precautionary principle" in the context of Western European environmental governance initiatives in the later twentieth century (Aradau and Van Munster 2007, 101–2; Ewald 2002, 283). Thus, while a number of (occasionally contradictory) versions of that idea have subsequently emerged in various contexts (Sunstein 2007, 123), its most commonly cited articulation is that provided by the *Declaration on Environment and Development*, drafted at the 1992 United Nations "Earth Summit" in Rio de Janeiro. The relevant clause asserts that "where there are threats of serious or irreversible damage, lack of scientific certainty shall not be used as a reason for postponing cost-effective measures to prevent environmental degradation" (United Nations 1992). While referring specifically to the environmental context, the underlying logic of precaution as a more radical modification of the general logic of risk is clearly embedded in this formulation, as the principle affirms that in the face of significant uncertainty regarding future disasters, preventative action in relation to such worst-case scenarios is both practically possible and ethically necessary. It also implies that all steps must be taken to avoid any course of action that harbors the potential to cause such catastrophic irruptions itself. According to the logic of precaution, in other words, radical "uncertainty is not an excuse" either for inaction regarding the prevention of future catastrophe or for a failure to exercise adequately imaginative foresight when deciding upon a course of action (Ewald 2002, 288). Precaution attempts to resolve the epistemological dilemma posed by the specter of a decision made in the face of radical, incalculable uncertainty by positing that such a decision can, in fact, be made. However, it also implies that this decision can only be legitimate if it is undertaken in reference to worst-case scenarios—irrespective of their connection to the available catalogue of risk knowledge (Ewald 2002, 286, 292; Furedi 2009; Sunstein 2007). From the precautionary perspective, therefore, anticipatory decisions remain possible in the face of a radical uncertainty that transcends the calculative rationality of risk, but only if *all* potentially catastrophic futures—even if their probability is estimated to be less than one percent—are adequately considered and taken into account (Sunstein 2007, 136).

In this sense, the concept of precaution and its underlying decisional logic represent a significant break from established modes of legitimate decision-making more generally, and the logic of risk from which it is conceptually derived in particular, in that it effectively severs the link between empirically based knowledge and an actionable decision (Ewald 2002, 288; Aradau and Van Munster 2008, 32). Indeed, precisely by virtue of its constitutive link to extreme, incalculable uncertainty, the logic of precaution requires that decisions be undertaken primarily on the basis of conjecture, speculation, and imagination, rather than empirically verifiable fact, or even a collection of inevitably partial or fragmentary risk knowledge. While perhaps unavoidable in a context of radical uncertainty, this ruptures in important ways with prevailing logics of rational action. By requiring that one "anticipate what one does not know" in reference to an imagined worst-case scenario, the decision-making process inevitably requires actors "to take into account doubtful hypotheses and simple suspicions . . . to take the most far-fetched forecasts seriously . . . whether true or false" (Ewald 2002, 288). Where the logic of precaution is invoked, in other words, all contingencies, regardless of their plausibility, must be considered, since what knowledge is available can never definitively exclude them as possibilities *a priori*.

Now, this does not imply that all potential contingencies will be granted an equal degree of attention or depth of response in any particular context, as the logic of precaution in the context of societal governance also involves decisions regarding which potentialities will in fact be anticipatorily acted upon. Moreover, those contingencies that are considered will themselves be selected in accordance with broader discourses of danger or threat—a process that is itself inexorably political and bound up with prevailing relations of identity and power (Campbell 1998). It is important to note that the logic of precaution still allows for a significant degree of agency among those actors structuring their behavior in accordance with it, and is thus also highly political. However, the key point for present purposes is that a precautionary logic of governance enacts an epistemic climate in which imagination trumps practical knowledge and, as Ewald, invoking Descartes, puts it, "invites one to make the most deceptive malicious demon one's closest companion" (2002, 289).

There are thus a number of potential pitfalls inherent in any political rationality rooted in logic of precaution. Perhaps the most significant is that a precautionary approach "reintroduces a pure logic of decision" into the realm of the political (Ewald 2002, 298). What this means is that, by severing of the decision from verifiable knowledge, the logic of precaution confers upon the decider a great deal more power, since it is their accepted construction of the future—rather than any established body of risk knowledge—that is the primary determinant of subsequent action (Ewald 2002, 298; 1993, 224). This has the effect of rendering precautionary decisions exceedingly arbitrary while also diminishing the accountability of the decision-maker, since the latter can always claim to have made the decision in a context of radical uncertainty in which no expectation of accuracy could be guaranteed (Aradau and van Munster 2008, 35). From this perspective, the potential for a descent into a condition in which groups or individuals arbitrarily constructed as threatening are anticipatorily targeted for violent interventions by relevant authorities is drastically increased with the adoption of a precautionary approach to governance, since the latter transfers the burden of proof to those cast as potentially threatening (Aradau and van Munster 2009; Diprose et al. 2008). Indeed, in the face of such pure logic of decision, it is up to those accused to prove otherwise, since the imagined threats they are deemed to embody have not yet materialized, and while they may never do so, they also cannot be ruled out (Diprose et al. 2008, 274; Stern and Wiener 2006, 398). The potentially deleterious implications for ostensibly liberal democratic societies are clear, as the inherently arbitrary nature of a politics of precaution renders it a potentially problematic and anti-democratic political tool—points that will be explored at length in later chapters.

Despite these potentially harmful implications for the exercise of political power, the logic of precaution offers a compelling rationality for both acting in the face of radical uncertainty and recapturing a degree of control over temporal contingency. As such, it has become an important aspect of governance strategies that operate well beyond the sphere of environmental policy from which it emerged, as Ewald's two features of a condition amenable to precautionary governance—"scientific uncertainty on the one hand, and the possibility of serious and irreversible damage on the other"—are increasingly perceived to obtain across myriad issue areas in contemporary human affairs (2002, 283–84). Broadly speaking, the modes of anticipatory governance that have been developed in response to the exigencies of late modernity are often more heavily influenced by the precautionary offshoot of the logic of risk than by the latter. This has led some scholars to describe the governmental rationalities currently at work in such radically uncertain political spheres as the global War on Terror as a "logic of precautionary risk"—something of a hybrid rationality that draws political legitimacy from claims of association with the scientifically calculative logic of risk, while operating primarily through the more speculatively arbitrary decisional logic of precaution (see Aradau and van Munster 2011, 2007, 2008).

It is these sorts of future-oriented, anticipatory political strategies—ultimately rooted in the logic of risk but operationally closer to the logic of precaution—that have increasingly come to fill the void created by the growing untenability of taking for granted sovereignty's spatialized resolution to the problem of temporal contingency and similar disruptions to established logics of governance and (temporal) control. In other words, the "reconfigur[ation of] the politics of space into a politics of time" that has accompanied the emergence of contemporary "risk society" can be understood as a reconfiguration of the spatialized politics of (sovereign) states into a temporalized politics of (precautionary) risk (Beck 2005; Kessler 2011, 2181). Thus, while the logic of sovereignty has long constituted a dominant political response to the problem of taming temporal contingency, the perceived radicality of the uncertainty we now face combined with the resultantly enhanced precarity of our phenomenological experience of time and futurity, has led to a significant shift in the way sovereignty is performed and enacted. Indeed, sovereignty is a historically contingent concept that must be continually reproduced through interventionary mechanisms of governance (Weber 1995), and contemporary challenges to its logic of temporal control have affected the way it is performatively enacted, such that the exercise of sovereign authority is becoming more overtly oriented toward imperatives of temporal control rather than merely spatial governance—a point evidenced by the widespread proliferation of governance strategies and practices explicitly rooted in the logics of risk and precaution. Understanding what is at stake in this ongoing reorientation of the political from the spatial to the temporal, from the reactive to the anticipatory, requires coming to terms with the fundamental implications of such a shift in political imagination for the organization and exercise of political power. This is the animating concern of the remainder of this book, as the focus narrows to one particular issue area in which the emergence of such anticipatory political rationalities has been particularly conspicuous: the post-9/11 global governance of (in)security.

NOTES

1. At this point, it is worth clarifying that my concern here is with the notion of *state* sovereignty—referring to the political principle most basically characterized by the establishment of a supreme political authority within an explicitly bounded territorial space. However, I follow Robert Jackson (2007) and use the shorter term "sovereignty" to signify this meaning in the present context. I am not unaware of the potential problems with such a move—as highlighted by the work of Prokhovnik (2007) among others. However, as R. B. J. Walker asserts, "As far as most contemporary forms of political analysis are concerned, sovereignty is quite obviously a shorthand for state sovereignty" (2010, 100). As such, I will employ these terms interchangeably in the context of this discussion. It should also be noted, however, that this orientation will change in subsequent chapters, as I shift to a broader understanding and use of the term sovereignty in recognition of the fact that, in the context of contemporary world politics, actors other than titular state executives often exercise powers that can be understood as sovereign within certain domains of action (see chapter 3).

2. Once again, this term is appropriated from Ancona et al. (2001, 645), in which the authors, albeit in the very different context of management theory, posit the analytical value of a "temporal lens [that] puts time . . . front and centre."

3. Although it should be noted that speculation *is* often marshalled as the primary basis for anticipatory action; however, such practices are more accurately associated with the idea of "precaution," on which more below.

Chapter Three

"To Kill Him Who Is Making Ready to Kill"

The Anticipatory Governance of (In)security

As indicated in the conclusion of the preceding chapter, much of the remainder of this book will explore the implications of the proliferation of anticipatory governance by focusing specifically upon how this trend has manifested in the context of (inter)national security. Such a focus is apposite, since the ongoing shift from a primarily spatially focused, reactive orientation to a temporally inflected, future-oriented concern with taming uncertainty through anticipatory interventions is a major feature of the contemporary global security landscape. Indeed, the international security environment has in recent years taken on many of the characteristics described by the narrative of diminishing temporal control discussed in the previous chapter, in that "security issues have increasingly been defined in terms of uncertain, potentially catastrophic threats" (Aalberts and Werner 2011, 2191). The post-9/11 preoccupation with transnational terrorism is at the root of this development, as terrorism operates largely beyond the familiar territorial logic of state sovereignty (Aalberts and Werner 2011, 2188; Kessler 2011, 2168) and has accordingly been problematized as a novel type threat that is uniquely "unpredictable in occurrence, characteristics, and effects" (Anderson 2010b, 228). The result is that security strategies are no longer conceived primarily in the spatialized terms of traditional "high politics"—whereby the defense of the sovereign state's territorial integrity against armed incursions from the outside is prioritized. Rather, the ostensibly ever-present specter of *potential* catastrophe—epitomized by the proverbial "next terrorist attack" (Aradau and van Munster 2011)—has inscribed the radically uncertain, potentially catastrophic future itself as the primary threat against which security action must be taken. This has led the very idea of "security" to become framed in explicitly temporal terms and equated with the taming of this future's putatively dangerous contingencies through targeted anticipatory action in the present. A key consequence has been the widespread proliferation of security strategies premised upon the notion that adequately securing a particular space now requires a measure of sovereign control over the vicissitudes of time—a direct inversion of the logic of sovereignty's spatialized solution to the problem of temporal contingency discussed in the preceding chapter. This is illustrated, for example, by the way the explicitly spatial idea of "Homeland Security" is now directly associated with the overtly temporal activity of preempting future catastrophic events through doctrines, strategies, and practices rooted in logics of risk (US Department of Homeland Security 2011). In short, then, this marked "shift from a reactive security politics to a precautionary politics" (Aalberts and Rijsdik 2011, 2157) constitutes perhaps the defining element of the post-9/11 global security climate, and also represents a clear illustration of the inversion of the erstwhile primacy of the spatial over the temporal described by Rosa and Scheuerman (2009, 10). Indeed, these developments have placed the problem of temporal control at the very core of the contemporary global politics of (in)security.

It is for these reasons that focusing specifically on the governance of (in)security can provide significant insight into what is at stake in the ongoing proliferation of anticipatory governance more broadly. This study's interest in the latter question will thus be pursued through an in-depth interrogation of how the problems of temporal contingency and radical uncertainty have been both prioritized within the global security imagination and responded to through the development of temporally inflected, future-oriented security strategies premised upon what can be best termed a "logic of preemption" (Anderson 2010a, 790).[1] The remaining chapters will be devoted to developing a comprehensive conceptual account and attendant critique of such strategies—which I will collectively refer to as "preemptive security."[2]

The present chapter begins this process by discussing what preemptive security can be understood to mean and establishing some conceptual and analytical parameters within which its critical interrogation can take place. Toward this end, the first section places the contemporary politics of preemptive security in historical and conceptual context to illustrate how it can be considered something of a unique political phenomenon. This is done by contrasting how preemption has been conceptualized post-9/11 to the treatment of the idea in the established canon of international law—which has heretofore been the most prominent context in which the anticipatory notion of preemption has been applied to problems of (inter)national security. In this respect, it is argued that the post-9/11 conceptualization of preemption is qualitatively different from its more conventional application in international law, as it casts radical uncertainty not as an *impediment to* anticipatory action but rather as the *basis for* it. Indeed, it is this effectively precautionary aspect of the post-9/11 conceptualization of preemption as a security logic that renders it historically and conceptually unique, and thus is in need of the sort of critical scrutiny to which it will be subjected in subsequent chapters.

However, before such an investigation can be pursued further, a series of analytical parameters and conceptual and terminological clarifications must be established to facilitate a better understanding of the subsequent arguments. This is the task of this chapter's second section, which begins by considering the question of ontology and contending that the notion of preemptive security (and by extension, the notion of anticipatory governance) can be best understood as a "political rationality" in the manner conceptualized by Rose and Miller (1992). I then consider how the relationship between agency and structure operates within the context of preemptive security rationalities before moving on to outline the particular understanding of sovereignty that is required to adequately capture these dynamics and which will be used throughout the remainder of the book. The chapter concludes by speaking to the scope of the analysis, with particular attention paid to the issue of Eurocentrism as it relates to the sort of general conceptual argumentation to which this book aspires. Having considered these questions, it will be possible to proceed with the core of the book's arguments and analysis in part II.

PREEMPTION AND (IN)SECURITY: INTERNATIONAL LAW AND THE WAR ON TERROR

While a convincing case can be made that a "shift from responding to past events to preventing future harms" constitutes "one of the most significant but unnoticed trends in the world today" (Dershowitz 2006, 7), the political imperative to control the unfolding of time in such a way is no new invention of our contemporary "risk society." Indeed, as has already been discussed at several points, the idea of exerting agentic influence upon the unfolding of the future through prudent interventions in the present has been a crucial, if often implicit, aspect of political thought since long before Machiavelli identified his idealized prince's *virtú* with the mastering of *fortuna* (Adam 2004; Bernstein 1998; Pocock 2003). It is thus important to help place the contemporary proliferation of anticipatory governance in context by considering the conceptual history of such ideas. While a comprehensive exploration is beyond the scope of this chapter, due attention can still be granted to this history in a manner more directly related to this book's aims by narrowing the focus to the conceptualization of preemption within discourses specifically related to problems of (inter)national security.

In this respect, it is useful to contrast the way preemption has been applied to questions of (inter)national security in the contemporary context of the War on Terror with the more traditional applications articulated in the canon of international law. Indeed, jurists and publicists have grappled with some variation of preemption since early modern thinkers began to seriously consider the idea of international law itself, and as legal theorizing about the use of force among states evolved, a particular conceptual formulation of preemption emerged that has become embedded in both customary and positive international law on the subject. Historically, this represents the primary discursive channel through which the notion of anticipatory action has been applied to questions of (inter)national security, and so the following discussion begins by tracing how preemption has been conceptualized in juridical treatments of the use of force. The main argument here is that international legal discourse articulates a conceptualization of preemption that, while still concerned with governing an inexorably *un*certain future, is nonetheless premised upon a high threshold of certainty as the basis for action. In this sense, the more traditional conceptualization of preemption found in the canon of international law is somewhat analogous to the idea of "risk" discussed in the preceding chapter.

This is a key point because, as the subsequent section argues, the notion of preemption that has been (re-)introduced into the security imaginary through the prosecution of the global War on Terror departs from this conceptualization by inscribing radical *un*certainty as the impetus for action. In other words, under the formulation of preemption that informs post-9/11 responses to transnational terrorism, a lack of certainty is not, as in the international legal context, conceptualized as a *barrier to* anticipatory action; rather, it is the very *basis for it* and, as such, moves away from an analogy to the logic of risk and toward the logic of precaution. The conceptualization of preemption that has developed in the post-9/11 era is thus qualitatively different from the way it has previously been applied to questions of (inter)national security; as such, its proliferation as a dominant paradigm of (in)security governance has serious implications for such fundamental sociopolitical questions as what "security" itself can be understood to mean, and how political subjectivity is experienced in the liberal democratic societies at the forefront of implementing anticipatory security strategies. Part II of this book examines such concerns in greater depth, in order to develop critical insights into logics of anticipatory governance more broadly; however, before proceeding in this direction, the qualitative shift in the conceptualization and application of preemption in the security context must be considered in more detail, beginning with an account of its articulation in the discourse of international law.

Preemption in International Law[3]

The idea of preemption has long been central to juridical treatments of the use of force in international affairs. This is because any comprehensive consideration of legitimate violence must address the difficult problem of balancing a putatively innate right to self-defense against the possibility of sanctioning aggression under the guise of anticipatory action (Kurtulus 2007, 222). The problem of preemption thus remains a major point of contention in international legal circles, with most jurists generally supporting some provision for preemptive action against an other in the face of manifestly imminent danger to the self (Dershowitz 2006). This position justifying anticipatory intervention when an actor is all but certain of imminent harm if no action is taken has a considerable historical lineage in the annals of international legal thought (Arend 2003). Indeed, the most influential early international legal thinkers included explicit formulations of preemption along these lines in their treatments of the laws of war, which have in turn become embedded in current customary and positive international legal frameworks. To put this another way, a central characteristic of preemption's conceptualization in international legal discourse is a high threshold of certainty vis-à-vis the future event to be preempted.

To elaborate on this point, it is useful to first consider the work of Hugo Grotius—often described as the "father" of international law—since he developed one of the more comprehensive engagements with the question of preemption among the proto-international jurists of early modernity and his arguments have been among the most influential in shaping subsequent juridical understandings of the idea (Clough and Stiltner 2007; Kegley and Raymond 2004, 41). In his magnum opus, *On the Law of War and Peace*, Grotius clearly affirms a right "to kill him who is making ready to kill," which he traces to universal precepts of Natural Law regarding self-defense (Grotius 2012 [1625], 82, 84). However, as Clough and Stiltner emphasize, Grotius was explicitly "concerned to ensure" that his conceptualization of preemption "was not interpreted too broadly"; thus, he also made the point that preemptive action required "moral certainty" regarding both the capability and the intention of the adversary (2007, 259, 253). Indeed, while he concedes "that when an assailant seizes any weapon with an apparent intention to kill, I have a right to anticipate and prevent the danger," he subsequently asserts that it would be gravely "mistaken" and "mislead[ing]" to "maintain that *any degree of fear* ought to be ground for killing another to prevent his *supposed* intention" (Grotius 2001 [1625], 64; beginning here, all emphases in quotations from Grotius are added). For Grotius, therefore, the normative argument for preemption is based upon a very narrow conceptualization of the idea that is characterized by two core components: a limited temporal horizon, whereby "the danger . . . must be immediate and imminent in point of time," and a very high threshold of certainty, whereby anticipatory action can only be taken "if it is *altogether certain* that the danger cannot be otherwise avoided" (2012 [1625], 83–84).

This latter point in particular is made in several places in *On the Law of War and Peace*, where Grotius's lines of argument imply that the concept of preemption as he understands it refers to a circumstance in which the actor in question possesses at least near-certain knowledge of an adversary's future actions and thus takes anticipatory action. For example, in a later chapter on the wider causes of war, he alludes to his understanding of preemption as self-defense with the following measures: "In order that self-defense may be lawful, it must be necessary; and it is not necessary *unless we are certain*, not only regarding the power of our neighbour but also *regarding his intention*" (Grotius 2012 [1625], 302). Crucially, Grotius takes this point further by also explicitly rejecting *uncertainty* as the basis for preemptive action. Instead, he argues that because "complete security is never guaranteed to us . . . [f]or protection *against uncertain fears*, we must rely on Divine Providence and on a wariness free from reproach, *not on force*" (2012, 90). The conceptualization of preemption that informs Grotius's theory of international law thus refers only to the taking of anticipatory action in the face of a temporally proximate threat about which the actor in question possesses near-absolute certainty. Indeed, anticipatory action that does not meet these criteria is neither ethically legitimate for Grotius nor—and this is the key point—conceptually congruent with the *idea* of preemption as he understands it.

I have used Grotius as an exemplar of the early modern jurists who first addressed the concept of preemption in depth and thus bequeathed a particular conceptualization thereof to the subsequent development of international law. However, it should be noted that some of his contemporaries and immediate successors did conceptualize a more expansive understanding of preemption that included a lower threshold of certainty. For example, Samuel von Pufendorf asserted in *On the Duty of Man and Citizen According to Natural Law* that one should not be forced "to suffer the first blow" before acting in self-defense, and should consider anticipatory action even if the adversary "has not fully revealed his design" (Pufendorf 1991 [1673], 50; see also Dershowitz 2006, 64–65). Nevertheless, as Clough and Stiltner point out, subsequent treatments of preemption in both customary and positive international law are much closer to the more conservative conceptualization articulated by Grotius (2007, 261, n. 20).

This becomes clear when considering the famous *Caroline* case of 1837, which remains to this day the touchstone around which preemption is understood in international legal terms. This case concerned an obscure Anglo-American border incident in which British agents in Upper Canada under the command of Sir Allan MacNab of Hamilton crossed into US territory and preemptively destroyed an American ship (the titular *Caroline*), which was being prepared by Canadian republican rebels under William Lyon Mackenzie for use in a forthcoming raid across the Niagara River. One American was killed and the ship was dramatically set ablaze and sent plunging over Niagara Falls, causing a significant uproar in US political and public opinion circles (Nichols 2008, 2). British authorities responded by claiming that the anticipatory act had been undertaken in self-defense, and in the course of the subsequent negotiations to resolve the dispute, US Secretary of State Daniel Webster provided what is still recognized as the authoritative statement on the status of preemption in customary international law. Webster famously acknowledged a right to preemptive action only under strictly limited circumstances, in which the intervening actor could demonstrate "a necessity of self-defense, instant, overwhelming, leaving no choice of means and no moment of deliberation" (quoted in Kindred and Sanders 2006, 1138). These normative criteria suggest a conceptual understanding of preemption premised on a similarly high threshold of certainty as that articulated by Grotius. Indeed, reading the text of Webster's statement closely, anticipatory action can only constitute the sort of unambiguous "necessity of self-defense" he specifies if the preempting actor is entirely certain of an adversary's impending attack (Arend 2003, 90–91).

While the *Caroline* case remains in the realm of custom, Webster's statement is generally understood as the *locus classicus* of preemption in international law more broadly, as its general formulation has also been incorporated into positive international law (Kurtulus 2007, 222; Walzer 1977, 74–77). Though no explicit mention is made of preemption in the United Nations (UN) Charter,[4] other official UN statements on the subject are generally congruent with Grotius and Webster's conceptualizations. For instance, the 2004 report by the UN's High Level Panel on Threats, Challenges, and Change reads both the theoretical tradition exemplified by Grotius and the customary tradition of the *Caroline* case against the Charter's general proscription on the use of force, concluding that "a threatened state, according to long established international law, can take military action as long as the threatened attack is imminent, no other means would deflect it, and the action is proportionate" (United Nations 2004, 54). Again, reading this text closely, only a very high level of certainty regarding the nature and imminence of the threat could support a conclusion that "no other means would deflect it." This suggests a clear conceptual thread running through the various discursive components of the international legal paradigm regarding the conceptualization of preemption, and a core element thereof is the requirement of near certainty regarding the time, place, and nature of the threat to be countered by anticipatory action (Arend 2003, 96).

Before concluding this section, a note on the preemption/prevention distinction is required. In international legal discourse, anticipatory action that does not fit the limited criteria discussed thus far is generally considered normatively illegitimate and most often described as "preventive" rather than "preemptive" (Elshtain 2013, 17–21; Kegley and Raymond 2004, 41). While the ostensible difference between these two terms has long been a point of contention in juridical and philosophical discussions of the use of force, I have avoided engaging with the debate over the distinction between the two here. This was done for two reasons. First, the preemption/prevention distinction is not as stable as is often implied, as the precise point at which the first segues into the second has never been made clear, creating much terminological slippage and semantic confusion in discussions of anticipatory action in international affairs (Fletcher and Ohlin 2008, 162). Indeed, as former Australian Foreign Minister Gareth Evans usefully points out, the English language is unique in making so precise a linguistic distinction between these two variations on the theme of anticipatory political action (Evans 2004, 65). Similarly, and secondly, the putative distinction between preemption and prevention is more temporal than evidentiary—meaning that it turns primarily on the proximity of the threat in time rather than the level of certainty regarding the existence and nature thereof (Doyle 2006, 55; Elshtain 2013, 18–19). Because at this point I am primarily concerned with comparatively examining the latter aspect of preemption across discursive contexts, the distinction between the two is of largely peripheral relevance to the aims of this discussion.

In light of these considerations, therefore, how should the conceptualization of preemption be understood in the context of international legal discourse? The above discussion suggests the key point in this regard is that the concept of preemption in international law is premised upon both a short temporal horizon—meaning that the impending harm must be clearly imminent rather than merely possible at some future point—and a high threshold of certainty—meaning that there must be no reasonable doubt as to this imminence and the character of the threat (Arend 2003, 96). This latter point is of particular importance for present purposes, and we have seen that the foundational theorizing of Grotius, the necessity criteria associated with the *Caroline* case, and the latter's affirmation in the UN's High Level Panel report together indicate that the international legal conceptualization of preemption describes the taking of anticipatory action on the basis of a high degree of certainty that a failure to do so would have destructive consequences in the immediate future. Of course, as with any anticipatory decision, a residual degree of uncertainty remains a defining feature, since the type of action envisioned by this articulation of preemption still takes place before any such consequences practically obtain. However, an *indefinite* temporal horizon coupled with a condition of significant *uncertainty* as to the threat to be acted upon are fundamentally incompatible with the conceptualization of preemption under the international legal framework. When mapped on to the discussion of the previous chapter, therefore, the discourse of international law conceptualizes preemption as somewhat analogous to the logic of risk, but explicitly contrary to the logic of precaution.

International law has traditionally been the primary discursive context in which the notion of preemption has been applied to questions of (inter)national security. Thus, when analysts or policymakers use the term "preemption" with respect to such questions, they are most often referring to action conforming to the general conceptual parameters outlined so far. Moreover, returning to the wider themes of sovereignty, temporality, and futurity, the international legal conceptualization of preemption appears broadly congruent with the modern understanding of political temporality that was discussed in the previous chapter, in which the spatial logic of sovereignty is presumed to adequately govern time and its contingencies. Indeed, because it envisions a very limited scope for preemptive action—which is taken only with respect to an immediately proximate future about which much certainty can be posited—this traditional conceptualization of preemption suggests a very limited need or capacity to mobilize sovereign power for the purposes of more direct control of temporal contingency. This in turn suggests that extant mechanisms of temporal control, such as the spatial logic of sovereignty, remain largely adequate and are in no pressing need of augmentation. Given the emerging perception, discussed in the previous chapter, that this is no longer the case, it should be no surprise that the post-9/11 (re-)articulation of preemption has moved away from the limited parameters of international legal discourse to embrace a more expansive understanding of anticipatory action premised upon the mobilization of sovereign power toward the active governance of contingency across a longer temporal horizon and in the face of much higher levels of uncertainty. This contemporary formulation of preemption—which has emerged primarily through developments in the post-9/11 response to the threat of transnational terrorism—thus diverges markedly from the framework of international law, representing a qualitative conceptual shift toward a logic of governance more firmly rooted in the notion of precaution rather than risk.

Preemption after 9/11

The concept of preemption is ultimately about the interaction between present and future; as such, the post-9/11 articulation thereof can only be understood if it is properly situated within the broader discursive climate created by shifts in the global security imagination that have altered the way in which present and future are understood. These shifts are characterized by the emergence of an account of the *present* as inherently precarious and beset by a temporality of "extreme uncertainty" (de Goede and Randalls 2009, 859), "radical contingency" (Dillon 2008, 315), or some other such term (Beck 2008). This narrative applies the broader societal arguments discussed in the preceding chapter specifically to the realm of (inter)national security, asserting that the contemporary global security climate is characterized by the breakdown of established certainties in the face of new threats and hazards—primarily transnational terrorism—which existing techniques of governance are incapable of adequately managing (Anderson 2010a, 781; Beck 2002; Kessler and Daase 2008). The result is that "decision-makers are simply no longer able to guarantee predictability, security, and control" to the extent that was once believed possible. As a corollary to this reading of the present, the *future* is increasingly understood in terms of not just radical uncertainty but also "expected and undeniable catastrophe," typified by the proverbial "next terrorist attack" (Aradau and van Munster 2008, 23). Moreover, as a direct consequence of the radical uncertainty ostensibly prevailing in the lived present, the degree of imminence and the precise moment of catastrophic irruption remain perpetually unknown. In other words, the future is conceptualized as embodying an impending catastrophe that is both inevitable—in the sense that it *will* occur at some point—and unknowable—in the sense that we can never be certain *when* it will take place (Anderson 2010a, 779–80; Elmer and Opel 2006, 481). In contemporary global security imaginaries, therefore, the future is increasingly problematized as "a disruptive surprise"—inevitably disruptive in its effects, and necessarily surprising in its unfolding (Anderson 2010a, 791; Opitz and Tellmann 2015, 107).

The sociopolitical ramifications of this (re)framing of present and future in the security context have been quite significant. On the question of human subjectivity, for instance, a number of scholars have noted that the corresponding discourses of radical uncertainty and inevitable catastrophe have coalesced to foment an increasingly "neurotic" worldview among citizenries—particularly in the West—in which the condition of existence is characterized by an enduring fear that the next disaster may be just around the corner (Anderson 2007, 159; de Goede 2008, 161; de Goede and Randalls 2009, 871; Isin 2004). In more practical political terms, the question of "security" has, arguably, come to trump all others in the realm of societal governance. As Agamben (2002) puts it, "security imposes itself as the basic principle of state activity," as publics accept that catastrophe is inevitably coming; that any certainty as to the moment of its emergence or the scale of the devastation remains beyond our epistemic or actuarial capacities to grasp, and thus we acquiesce to attempts by sovereign authorities to confront this condition through "exceptional" means.[5] The result is the broader emergence of what Paul Virilio has pointedly termed "a culture of the imminence of disaster" (2010, 7), in which "the threat is known to have the ontological status of indeterminate potentiality" (Massumi 2007, 8). In other words, to paraphrase Donald Rumsfeld's now-folkloric idiom, the current global security climate—and thus arguably the contemporary human condition, particularly as experienced in the polities of the West—is haunted by the spectre of a catastrophic unknown that is nevertheless known to be lurking in the depths of the future.

Importantly, this apparently perpetual imminence of catastrophe has not led to an introspective embrace of contingency through a resigned "recognition of the fragility of modern life" (de Goede and Randalls 2009, 867; Fierke 2007, 203). Rather, these conditions have led to the emergence of a broader security ethos premised on confronting radical uncertainty by actively controlling the unfolding of the future through anticipatory interventions in the present. Typified by former US Vice President Dick Cheney's assertion regarding the purported threat posed by Saddam Hussein's Iraq that "the risks of inaction are far greater than the risk of action" (quoted in Daase and Kessler 2007, 426), such an outlook implies the necessity of taking action irrespective of the existence of an established base of verifiable knowledge upon which to make actionable decisions—that is, an effectively precautionary posture (Aradau and van Munster 2007, 101; Elmer and Opel 2008, 24). Put another way, radical *un*certainty about both the precise nature of the threat and the moment at which it is likely to emerge does not serve as an *impediment to* anticipatory action; it instead provides the *very impetus* for it, since the putative uncertainty of the present and the concomitant potential of future catastrophe imply that "[w]ithout some form of action, a threshold will be crossed and a disastrous future will come about" (Anderson 2010a,

780). In practice, this places governing authorities "in the uncomfortable position of having to take drastic action in the face of an inescapably elusive, uncertain threat" (Cooper 2006, 119). The result is a politics of (in)security in which "the absence of specific evidence serves as justification for action" (Elmer and Opel 2006, 481), and "cognitive uncertainty . . . [is] not seen as a problem to be solved, but as the justification for . . . action" (Daase and Kessler 2007, 426). Thus, in the words of a Brookings Institute policy paper describing the American approach to governing terrorism, the post-9/11 understanding of preemption "is not limited to the traditional definition of preemption—striking an enemy as it prepares an attack—but also includes . . . striking an enemy in the absence of specific evidence of a coming attack" (Elmer and Opel 2008, 23–24; O'Hanlon et al. 2002, 3).

In this respect, established mechanisms, techniques, and logics of (in)security governance nominally premised upon extensive data collection and in-depth deliberation are rendered inadequate, and the development of strategies premised upon anticipatory action rooted primarily in conjecture and speculation emerge as the only viable option. While this point will be discussed at greater length in later chapters, it can be clearly seen in the Obama administration's use of "predictive assessments about potential threats" as the basis for the inclusion of individuals on so-called no-fly lists and other associated "watchlisting" initiatives integral to its ongoing prosecution of the War on Terror (Ackerman 2015). For example, in a May 2015 federal court proceeding relating to a case brought against the US government by the American Civil Liberties Union, two officials from the Justice Department described the watchlisting program's goal as "identifying individuals who may be a threat to civil aviation and national security" through "predictive judgement[s] intended to prevent future acts of terrorism in an uncertain context" (quoted in Ackerman 2015). As a *Guardian* report on the proceeding describes it, decisions to place a particular individual on a no-fly list are thus "based on predicting crimes rather than on records of demonstrated offenses." Despite a Federal Bureau of Investigation (FBI) counterterrorism official telling the court that "mere guesses or 'hunches' . . . are not sufficient" to place someone on a no-fly list, a former Central Intelligence Agency (CIA) counterterrorism analyst criticized this claim by pointing out to the court that "there is no indication that the government has assessed the scientific validity and reliability of its predictive judgments"; as such, these judgments in practice do in fact "amount to little more than the 'guesses' or 'hunches' that [the FBI official] says are not sufficient" (quoted in Ackerman 2015). Looking beyond the questionable appeal to scientism in this context, the CIA analyst's criticism and the court proceedings more generally provide an instructive illustration of how the importance of taking action in the face of radical uncertainty characterizes the contemporary manifestation of preemption as related to questions of (inter)national security, and that this has significant implications for the way political decisions relating to such questions are made.

The key point, therefore, is that the post-9/11 formulation of preemption is qualitatively and historically unique in that it addresses the key questions of (un)certainty and imminence in a nearly inverse manner to the understanding of preemption articulated in the discourse of international law. Indeed, unlike the latter conceptualization, which, again, is premised upon a high degree of certainty regarding the threat in question, the post-9/11 formulation of preemption frames the mere *potential* of danger emerging *at some indefinite point* in the future as the basis for anticipatory intervention in the present. Accordingly, the temporal horizon against which the imminence of a threat (and thus the necessity of action) is measured is no longer limited to an immediate future whose contours can be ascertained with near certainty on the basis of available evidence in the present. Rather, this horizon is rendered effectively unlimited, while the threat to be addressed must be constructed through speculation since the extant condition of radical uncertainty precludes the acquisition of adequate "risk knowledge" about an event before it occurs. Thus, if anticipatory action is to be taken, the level of certainty that underpins it will not be nearly as high as it is under the articulation of preemption associated with international law. In fact, it must necessarily be low, since the very condition of radical uncertainty that such action is designed to mitigate effectively precludes the creation of a robust epistemic foundation upon which to decide if and how to act. A high level of certainty is thus neither required nor even achievable in this context, while the difficulty of acquiring verifiable knowledge regarding the threat only enhances, rather than diminishes, the necessity of acting anticipatorily.

A growing body of work has documented the widespread proliferation of practices embodying these characteristics in the context of the post-9/11 governance of (in)security (see, for example, Amoore 2014a; Anderson 2010a, 2010b; Cooper 2006; de Goede 2012, 2008; de Goede and Randalls 2009; Ehrenberg et al. 2010; Elmer and Opel 2006; Massumi 2007, 2005; Stockdale 2013). While addressing a wide range of issues, including but not limited to the indefinite detention (Butler 2006; Ericson 2008), extraordinary rendition (Mutimer 2007), and targeted killing (Grayson 2012; Leander 2011) of suspected terrorists; the widespread biometric monitoring of increasingly mobile populations (Amoore 2009; Muller 2010); the preemptive detention of refugees and asylum seekers (Isin and Rygiel 2007; L. Weber 2007); the preemptive freezing of monies and assets suspected of terrorist ties (de Goede 2012, 2011); and the so-called Bush Doctrine of preemptive war (Ehrenberg et al. 2010; C. Weber 2007), this literature has vividly illustrated both that the dominant trend in contemporary (in)security governance is a turn toward anticipatory action and that the formulation of preemption underpinning this trend is historically and conceptually unique in precisely the manner described here.

An alternatively useful way to understand these points is to recognize that the traditional formulation of preemption associated with international law remains an essentially *reactive* notion, since it involves acting on an acutely immediate future whose imminent emergence is all but certain. In other words, it is still about *responding to* a threat—albeit with perhaps a slight head start over a purely reactive posture (Doyle 2006, 15). As noted IR scholar Richard K. Betts puts it in his own description of preemptive war as an international legal concept, "Beating the enemy to the draw by striking before he launches his attack is reactive, even if it involves firing the first shot" (2003, 18). By contrast, the post-9/11 formulation of preemption—with its precautionary bent and concomitant inscription of radical uncertainty as the basis for action—represents an attempt to actively manipulate and assert control over the unfolding of the future, rather than more passively reacting to it. Understanding the post-9/11 conceptualization of preemption in this way further highlights its qualitatively unique character, since it shows that it is not merely a new iteration of "prevention" as understood and deemed illegitimate under international law. This is because prevention is *also* a reactive idea, in that it, too, ultimately implies a relatively high threshold of certainty prior to the taking of anticipatory action. Betts's description is again useful for illustrating this point, as he asserts that "the rationale for preventive war is that conflict with the adversary is so deep and unremitting that war is ultimately inevitable, on worse terms than at present" (Betts 2003, 18). This implies that "prevention" is distinguished from traditional preemption less by the substantive existence of the threat—about whose "deep and unremitting" nature the preventing actor is still quite certain—than by the temporal proximity of its transformation into an acute danger. As such, preventive action is still essentially a *reaction* to an ontologically tangible threat about which the actor possesses a high level of certainty.

Moreover, while closely related in most key respects, the broader concept of precaution discussed in the preceding chapter can also be differentiated from the specific post-9/11 iteration of preemption along similar lines. Indeed, some have argued that despite constituting a rationality for taking action in the face of incalculable uncertainty, "precaution . . . appears to be still grounded in the assumption that the threat can somehow be known or experienced." In other words, the threat's basic ontological presence is clearly established before action is taken, and anticipatory measures only "seek to act before an identified threat reaches a point of irreversible damage" (Anderson 2010a; Leese 2014, 498). This suggests that precaution is *also* ultimately a reactive idea, in that it involves proactively responding to a threat whose present acuity may be minimal, but whose existence can be verified in a convincingly objective manner.

By contrast, we have seen that preemption, as manifested in the post-9/11 security climate, thus "anchors itself 'unabashedly' in uncertainty" (de Goede, Simon, and Hoijtnik 2014, 413), functioning explicitly as a rationality for acting in the face of the perceived "objective uncertainty" created by the spectre of transnational terrorism in particular (Massumi 2007, 8). This formulation of preemption therefore refers to action taken even "before the formation and identification of a determinate threat" (Anderson 2010a, 792; Leese 2014, 498), and "valorizes the interval between now and the catastrophe to come as a domain of decision and action within this world" (Opitz and Tellmann 2015, 112). In practice, such action requires the creative imagining of a possible catastrophic future and the subsequent tailoring of anticipatory interventions to explicitly preclude that future from coming to pass. As Massumi puts it, "preemption actualizes the potential in a shape to which it hopes it can respond" (2007, 12). Thus, another way of understanding the unique character of the post-9/11 formulation of preemption is to recognize that, unlike the concepts of prevention or even precaution, it is not so much reactive as *productive*. Indeed, the ostensibly abyssal uncertainty against which action must be taken requires the imaginative *construction* of dangerous futures around which present anticipatory interventions are framed, and the carrying out of these interventions *actively creates* an alternative future from which the potential danger(s) in question are removed. It does not react to an impending future outside of its control; it actively *produces* its own future by anticipatorily foreclosing the catastrophic futures creatively imagined by its practitioners.

Another example from the ongoing War on Terror can help illustrate these points. On February 25, 2015, Abdurasul Juraboev, Akhror Saidakhmetov, and Abror Habibov—all recent immigrants from Central Asia residing in Brooklyn, New York—were arrested by the FBI and charged with "providing material support to a terrorist organization" (Federal Bureau of Investigation 2015; Santora and Clifford 2015). The development of the case against them followed a very common pattern among FBI counterterrorist operations, in which suspects are placed in contact with a paid informant who then induces the suspects to begin actively planning some sort of attack—activities for which they are subsequently arrested and charged (Santora and Clifford 2015). Journalistic reporting on this particular case has indicated that prior to the start of the FBI's investigation, the three men's only confirmed involvement in any terrorism-related activity consisted of posting sympathetic statements on a website affiliated with the so-called Islamic State in Iraq and the Levant (ISIL) (Hussain 2015). Indeed, according to the *New York Times*, the arresting authorities admitted that their "threats of violence had an 'aspirational' quality to them" and that there was "no indication that the suspects were close to staging an attack, large or small" at the time of their arrest. Not only was this clearly an anticipatory intervention by the FBI, but the latter's paid informant also appears to have been the driving force behind the plot that was preempted, in that "only after the introduction of the informant did any actual arrangements to commit a criminal act come into existence" (Santora and Clifford 2015). The FBI thus actively produced the very future threat against which it subsequently intervened preemptively.

It is important to note that this was not an isolated incident, as this technique of utilizing paid informants to goad suspects into committing terrorist-related inchoate offenses—often to the extent that the informants "actively develop the plot" and "provide the resources to carry it out"—has been a vital component of post-9/11 American domestic counterterror strategy (Human Rights Watch 2014; Bartosiewicz 2012).[6] These practices are further facilitated by the court system, which legitimates this sort of so-called preemptory prosecution through highly permissive readings of legal provisions against entrapment (Greenwald and Fishman 2015).[7] This is illustrated by the 2011 case of *United States v. Cromitie*, in which the presiding judge upheld the defendant's conviction and twenty-five-year sentence for several terrorism-related offenses despite writing in her ruling that "the Government indisputably manufactured the crimes of which the defendants stand convicted . . . invented all of the details of the scheme . . . [and] funded the entire project," while also forthrightly recognizing that the FBI's informant "was the prime mover and instigator of all the criminal activity that occurred" (*United States v. Cromitie*, 2011).

The key point for present purposes is that the proliferation of these particular methods—through which US authorities have effectively "created terrorists out of law-abiding individuals" (Human Rights Watch 2014, 22)—has corresponded precisely with a post-9/11 shift in the FBI's mandate that "fundamentally refocused the Bureau from its traditional law enforcement mission to gathering intelligence on *potential* threats" so as to enable it "to catch the terrorists *before* they can strike next" (Bartosiewicz 2012, emphasis added). In other words, the strategy of actively manufacturing terrorist plots that portend a potentially catastrophic future that can subsequently be disrupted through anticipatory action is a direct function of the FBI's operationalization of the preemptive ethos that has underpinned the wider post-9/11 governance of (in)security. This is congruent with the conceptual argument developed above that the imperative to act in a climate of radical uncertainty requires the active creation of potentially catastrophic futures to function as the basis for anticipatory action aimed at producing an alternative future that precludes such threats. Indeed, a counterterrorism strategy in which, as one prominent critic has put it, the FBI "does not disrupt planned domestic terror attacks but rather creates them, then publicly praises itself for stopping its own plots," is precisely the type of practice suggested by this logic, and provides a clear illustration of the contemporary formulation of preemption's overtly productive character (Greenwald and Fishman 2015).

A New Logic of Preemption?

To sum up the key points discussed thus far, the post-9/11 articulation of preemption in relation to (inter)national security—which has emerged out of attempts to confront the ostensibly unique security problem of transnational terrorism—all but inverts the core conceptual premises of the more traditional conceptualization of preemption articulated in the discourse of international law. Indeed, in the context of the global War on Terror, acting preemptively no longer denotes simply intervening anticipatorily to counteract an event whose nature and imminence are both known to a reasonably high degree. It instead refers to intervening in the present on the basis of what is imagined may happen at some point in an indefinite future, so as to thereby actively produce an alternative future in which this imagined catastrophe is necessarily precluded. In other words, the contemporary logic of preemption is premised upon taking anticipatory action not only in the face of but also on the basis of radical uncertainty. In contradistinction to the discourse of international law that conceptualizes preemption in a manner reminiscent of the logic of risk, the form of preemption that has emerged as the dominant conceptual underpinning of post-9/11 (in)security governance appears much closer to the more extreme logic of precaution—though, as discussed above, differing in important, even more radical ways (Aradau and van Munster 2007). Additionally, the specific practices of preemptive security examined in the literature cited earlier and illustrated by the examples discussed above demonstrate not only that the notion of preemption has undergone a significant conceptual shift in terms of its treatment of (un)certainty and imminence but also that its application to the realm of (inter)national security is now understood to include the anticipatory use of state power against particular individuals or groups, rather than simply other states (as in the international legal context). This represents a further conceptual modification, which extends the practical scope of preemption in the global security context in ways that exacerbate its implications for the exercise of sovereign power.

Finally, to connect this section's discussion to the themes of the preceding chapter, the key difference between the post-9/11 articulation of preemption and that found in the discourse of international law can also be understood in terms of the question of temporal governance. In this regard, the essentially reactive character of the traditional juridical conceptualization of preemption embodies a more passive logic of temporal control, since it effectively involves responding to the imminent unfolding of time. By contrast, the post-9/11 conceptualization of preemption is concerned with (re-)asserting a degree of agentic control over the vicissitudes of time through a continuously productive sequence of anticipatory interventions in the lived present. Understanding the distinction in this way hints at the potentially significant political implications of adopting such a preemptive approach to (in)security governance, as the shift from a subjective posture of relative passivity toward the future to one premised upon the active governance thereof implies potentially radical changes to the way security decisions are made in particular, and thus how political power is organized and exercised more generally. Unpacking these points is the core concern of part II of this book, but before proceeding in this direction, some analytical and conceptual parameters for doing so must be considered. This is the task of the next section.

CONCEPTUALIZING AND ANALYZING "PREEMPTIVE SECURITY"

Preemptive Security as a Political Rationality

The first point to be made can be considered ontological, as it concerns how the idea of anticipatory governance more generally—and preemptive security in particular—can be understood within the conceptual matrix of political analysis. In this regard, the idea of preemption in the sense used here can be best understood as what Nikolas Rose and Peter Miller, following Foucault, term a "political rationality" (Rose and Miller 1992; see also Aradau and van Munster 2007, 97). As discussed briefly in the introduction, this concept refers to a "discursive field within which the exercise of power is conceptualised," which combines "justifications for particular ways of exercising power by diverse authorities" with "notions of the appropriate forms, objects, and limits of politics, and conceptions of the proper distribution of such tasks" (Rose and Miller 1992, 175). Put differently, political rationalities are discursive constructs concerned with "the formulation and justification of idealised schemata for representing reality, analyzing it, and rectifying it" (178). A political rationality thus "problematizes" a certain aspect of the social world, and offers a framework through which "programmes of government" can be developed in response to the identified problem(s) (181–82).

Importantly, because they are concerned with both framing a social problem as in need of rectification and providing a governmental framework through which it can be addressed, "political rationalities have a characteristically moral form" that "consider[s] the ideals or principles to which government should be directed" (Rose and Miller 1992, 178–79). In other words, political rationalities are explicitly premised upon an underlying normative claim regarding the particular realm of human affairs with which they are concerned, as they both affirm that this area requires action and offer a program through which such action might be taken. Yet, because they imply such a practical dimension whose actualization will be subject to the inexorable vagaries of social context, political rationalities "do not have the systematic and closed character of disciplined bodies of theoretical discourse" (178). They thus do not imply a tightly bound logic within which there is no flexibility and that is not subject to situational variation upon its implementation. Rather, political rationalities merely possess certain "regularities," which translate into a general logic of praxis that guides the contextual political action that results. In other words, even though its implementation may not take the same form in every circumstance, a political rationality does nevertheless imply a general pattern of operation that can be treated as consistent for the purpose of analysis and critique. It is a useful conceptual label for the notion of anticipatory governance in the context of this study, since the aim is to subject the logic of preemptive security—as an archetypical representation of anticipatory governance—to precisely this sort of interrogation.

Indeed, the post-9/11 articulation of preemption discussed in the preceding section fits Rose and Miller's characterization quite well. Consider in this respect that, just as in Rose and Miller's formulation, the idea of preemptive security is premised upon a very particular representation of a social problem—namely, (in)security—which casts it in temporal terms by emphasizing radical uncertainty and the constant potentiality of catastrophe. As with a political rationality, it is premised upon a normative claim that actively taming this uncertainty should constitute a primary governmental imperative, and this moral imperative translates into a particular "program of government"—namely, the taking of anticipatory action in the present based on imagined future potentialities and the concomitant production of an alternative future. To be sure, the specific practices associated with such a program may vary contextually, perhaps quite significantly; however, this sort of variation is also consistent with Rose and Miller's idea of a political rationality. Finally, despite such potential variation, the underlying practical logic represented by this "programme of government"—namely, governance through anticipatory intervention—comprises a general pattern of operation, a set of consistent "regularities," that can be subjected to analysis and critique. The contemporary articulation of preemption as outlined in the preceding section fits Rose and Miller's characterization quite well and, as such, can be usefully conceptualized as a "political rationality." Moreover, by developing an in-depth conceptual account of preemptive security through a critical interrogation of its own internal logics, the remainder of this book pursues the sort of analysis to which Rose and Miller suggest political rationalities should be subjected. Thus, when preemption is referred to as a "rationality"—which it often is throughout the remaining chapters—it should be understood in this context and recognized that this term was deliberately chosen.

Agency-Structure and Sovereignty-Governmentality

Approaching these ontological questions in this way also helps address two additional issues that must be considered in order to properly frame the subsequent analysis. The first relates to the agency-structure question in the context of logics of preemptive security. While the arguments developed in part II do not deal directly with this fundamental meta-theoretical issue, its importance to all analysis of the social world requires that it be given some brief consideration at this juncture, as the subsequent arguments do presuppose a certain contextual understanding of it. In this respect, any conceptual engagement with the operational character of preemptive security must reckon, if only implicitly, with how relevant actors are both empowered through and constrained by their embedment within a preemptive security framework. The above description of preemption as a political rationality gestures toward a possible response to this question, in that political rationalities imply an overarching general framework through which action takes place while also allowing for significant variation at the specific level of practical implementation (Rose and Miller 1992, 178–79). Put differently, Rose and Miller's formulation implies an answer to the agency-structure question whereby agents operate within the broad confines of a particular discursive structure that directs their behavior at the general level, while also retaining significant autonomy with respect to the particular practices through which these overarching aims are implemented. In short, a discursively powerful overarching structure thus exists, but there is also much space for practical agency to be exercised within that structure.

Given that preemption can be considered a political rationality, this provides a helpful description of the interaction between agents and structures in the context of a preemptive security regime. Indeed, on the one hand, the relevant agents[8] are conditioned by the imperative to stop the proverbial "next attack" before it occurs through anticipatory intervention. Their behavior is thus ultimately directed by what Massumi (2007) terms "the primacy of preemption," wherein the goal of security action is, above all else, to ensure that imagined potential catastrophes do not, in fact, come to pass. This overarching imperative both guides and constrains the behavior of relevant agents. On the other hand, however, such agents are also granted a significant degree of autonomy with respect to practically implementing this broader imperative. Indeed, the practical exigencies of pursuing a preemptive security strategy with respect to terrorism, for example, require that considerable discretion be given to the various authorities tasked with carrying it out "on the ground." This is because the almost infinite number of potential catastrophes inhabiting a radically uncertain future cannot be adequately confronted by a highly centralized authority alone. Rather, a comprehensive preemptive security strategy requires a disaggregated framework in which subsidiary authorities possess significant levels of autonomy in terms of pursuing the overall goal of preempting potential threats. This point is highlighted by Aradau and van Munster (2007, 105), who describe how the anticipatory governance of terrorism has manifested as a "*dispositif* of risk" that "activates all the technologies imaginable in the face of uncertainty" and is thus operationalized through a broad array of agents acting to realize the overarching aims of anticipatory governance. Decision-makers retain a significant degree of practical agency vis-à-vis the discursive structure of preemption. Finally, the cycle is completed by the way their pursuit of a preemptive imperative in this fashion serves to performatively reaffirm the overarching structure as the ultimate frame within which (in)security governance is prosecuted.

The mutually constitutive character of this process might initially suggest a broadly structurationist account (Giddens 1984). Indeed, in accordance with the general tenets of structuration theory, the primary agents are conditioned by the discursively powerful notion of preemption; yet they retain a significant degree of autonomy whose exercise ultimately serves to reinscribe this notion as a powerful discursive structure. However, in lieu of a purely structurationist account, the relationship between agency and structure in the context of preemptive security is more accurately described by Judith Butler's concept of "sovereignty within the field of governmentality" (Butler 2006, 61). To be sure, this risks mixing theoretical metaphors, since Butler's conception was not developed as a direct response to the agency-structure question. Yet it nonetheless captures the dynamic just described quite well, and thus usefully illustrates the nature of the agency-structure relationship in the context of preemptive security. Further elaboration upon Butler's idea of sovereignty within governmentality will thus clarify the understanding of agency-structure that is presupposed in the remainder of this book. It will also provide a useful segue into another theoretical issue that must be addressed before proceeding: how the idea of sovereignty will be deployed and (re-)conceptualized in the subsequent chapters.

To understand Butler's idea, we must first consider Foucault's notion of "governmentality" (Foucault 1980, 1991). While it is a characteristically nuanced Foucauldian idiom, governmentality refers most basically to the "diffuse set of strategies and tactics" that characterize the governance of modern liberal societies (Butler 2006, 52; Foucault 1991). In this sense, Foucault developed the concept to capture the degree to which modern governance no longer emanates from a "unified sovereign subject," but instead operates diffusely through a variety of discursive and material institutions that collectively work to "dispose and order populations, and to produce and reproduce subjects, their practices, and their beliefs, in relation to specific policy aims" (Butler 2006, 52). Famously proclaiming the need to "cut off the king's head" in political theory, Foucault conceived of governmentality in opposition to the traditional unitary conception of sovereignty (1980, 121). In particular, he suggested that governmentality emerged as the dominant mode of governance as modern processes of liberalization "devitalized" the quasi-absolutist logic of traditional sovereignty, as represented in the theories of Bodin, Hobbes, and other prominent early modern figures in political thought (Butler 2006, 52–53).

On Butler's reading, however, this opposition between sovereignty and governmentality is merely an "analytic distinction" made by Foucault, and does not rule out the possibility "that these two forms of power can and do coexist in various ways" (Butler 2006, 54). Indeed, it is to precisely such a coexistence that she refers when developing the notion of "sovereignty within the field of governmentality." More specifically, in attempting to theorize the form(s) of governance that have emerged in post-9/11 America in particular, Butler formulates a broader account of political power that fuses both concepts in a manner that is quite relevant to the analysis of preemptive security. In this respect, Butler follows Foucault by arguing that the proliferation of governmentality is most visibly illustrated in the current moment by the ongoing "disarticualtion" of the state "into a set of administrative powers" (55). This phenomenon is particularly characterized by the downloading of what she terms "prerogative power" to unelected agents of the state—a managerialization of governance in which appointed "officials with no clear claim to legitimacy" are endowed with what amounts to sovereign power within a particular context (54). Conceptually, the key point is twofold. On the one hand, these officials still operate "within the constraints of governmentality," since "their power is delegated, and they do not control the aims that animate their actions" (62). Yet, on the other hand, they are also "delegated with the power to render unilateral decisions" within the particular contextual circumstances in which they are tasked to operate, so long as these decisions are ultimately geared toward the realization of the overarching aims of the governmental regime (56). Citing the array of officials comprising the vast American national security apparatus as archetypal examples, Butler compellingly argues that these decisions are effectively "sovereign," since they are in practice accountable to no higher authority than those unelected officials making them. As Butler puts it, "the power they wield . . . is a sovereign power" (59).

Such a paradigm of political power functions through a "resurgence of sovereignty in the midst of governmentality" (Butler 2006, 55). The overarching governmental aims are practically operationalized through the delegation of prerogative power to an array of bureaucratic officials, thus constructing them as effectively sovereign within their own limited domains. In this sense, "sovereign power emerges as the power of the managerial 'official'"—a figure that is "deployed by tactics of power [he or she] does not control" but who is also able to exercise decisional power in a way that "reanimate[s] a sovereignty that the governmentalized constellation of power appeared to have foreclosed" (65). Butler has famously labeled these figures "petty sovereigns," and it is important to recognize that this can refer to *any* agent endowed with such capacities within the broader governmental assemblage of the contemporary state—be it a border guard, an airport security official, a customs agent, an intelligence officer, a military field commander, or otherwise (56, 65).[9] It is through these figures that Butler's notion of "sovereignty within governmentality" is manifested.

To return to the particular concerns of this section, Butler's figure of the petty sovereign usefully illustrates the agency-structure relation I have ascribed to the logic of preemptive security, while also hinting at the conception of sovereignty that characterizes the latter's practical operation. With respect to the first of these, Butler describes petty sovereigns as "mobilized by aims and tactics of power they do not inaugurate or fully control," yet also "delegated with the power to render unilateral decisions, accountable to no law and without any legitimate authority" (2006, 56). In other words, they are ultimately constrained in their behavior by the prevailing structure of governmentality, since they act to uphold and operationalize an overarching set of policy aims; yet they nevertheless retain a significant degree of agentic autonomy within these constraints, to the extent that they can legitimately be called sovereign within their particular domain of action. As Butler succinctly puts it, "they are acted on, but they also act" (62). This depiction quite closely mirrors the relationship between agency and structure that I have posited as obtaining in the context of preemptive security. Again, in the latter case, the overarching governmental aims are the anticipatory governance of a potentially catastrophic future, and it is toward this ultimate end that those agents tasked with prosecuting a preemptive security mandate orient their decisions. However, given the possibly infinite potentialities implied by a radically uncertain future, this general set of governmental aims cannot dictate precisely what action should be taken in every possible circumstance that may arise. The relevant agents must thus be granted a significant degree of decisional autonomy within the broader climate of preemption if the latter is to be made actionable as a security rationality.

It is this productive tension between the "structure" of a preemptive imperative and the autonomous "agency" of those seeking to render it actionable that is key to the praxis of preemptive security. Accordingly, just as in Butler's formulation, relevant agents "do not fully control the aims that animate their actions"—in this case, the general idea of preemption—and are therefore constrained in their behavior by this overarching discursive/governmental structure (Butler 2006, 62). However, they also retain a significant degree of agency, since the general idea of preemption can only be translated into a functional security rationality by delegating prerogative powers to various agents tasked with making it so. Accordingly, Butler's words describing the status of the petty sovereign within the post-9/11 governmental structure of the United States can also be used to describe the agency-structure relationship that obtains in the context of preemptive security in the abstract: "their acts are clearly *conditioned*, but their acts are judgments that are nevertheless *unconditional*," since "they are constituted within the constraints of governmentality as those who will and do decide" (62, 65, emphasis original). A form of sovereignty is located "precisely within the agency of the functionary, and so, within the field of governmentality itself" (66). It is in this sense that Butler's notion of sovereignty within the field of governmentality is a useful analogue for the agency-structure relation that prevails in the context of a preemptive security rationality. Agents are constrained by the structure, but preserve a notable degree of agency within the broader parameters of that structure.

This similarity between Butler's conception of sovereignty within governmentality and the agency-structure relationship that obtains under a preemptive security rationality is no coincidence, however. There is an additional dimension to the connection between her paradigm of political power and the praxis of preemptive security, in that the types of agents and actors that she sees as embodying sovereignty within governmentality—the "petty sovereigns"—often also serve as the primary figures through which anticipatory security rationalities are practically operationalized. In other words, it is the wide array of bureaucratic officials, government functionaries, and executive branch surrogates to which Butler attaches the term "petty sovereign" that are often on the front line of implementing preemptive security strategies. This is because, as noted above, to be fully implemented such a strategy requires the mobilization of a broad, decentralized array of agents who operate in accordance with the general logic of preemption, but who are also granted a significant degree of autonomy in prosecuting this mandate "on the ground." In light of this point, a modification to the way the concept of sovereignty is understood in the context of this study is required.

To elaborate, the analysis has thus far taken the more traditional approach of deploying the term sovereignty more or less synonymously with the concept of state sovereignty, with the term sovereign power being implicitly equated with the prerogatives of the titular executive. However, a more expansive understanding based upon Butler's framework—which includes both the executive *and* the petty sovereign—better captures the way sovereignty and sovereign power operate in the relevant contexts. Indeed, the notion of a highly centralized governmental apparatus in which a wide array of functionaries simply follow the dictates of a central authority does not provide an adequately nuanced account of the way political power is exercised in the functioning of modern states, particularly in the security realm. Rather, sovereign power is diffusely constituted by the networked interaction of multiple nodes of authority, each exercising its own varied degree of agency vis-à-vis the center (Rose and Miller 1992; Slaughter 2004). Butler's notion of sovereignty within governmentality articulates this point in a compelling way. In addition to illustrating the sort of agency-structure relation that prevails under preemptive security rationalities, her framework also offers a sophisticated account of political power that is of particular relevance to the study thereof.

In the remainder of the book, I therefore move away from the narrow traditional conceptualization of sovereignty and toward one based upon Butler's ideas, in which sovereignty is understood to be exercised by any number of agents depending on the circumstances. Such a shift is necessary because the overlap between Butler's petty sovereigns and the agents often tasked with implementing a preemptive security rationality suggests that a limited, executive-only understanding of sovereignty cannot adequately capture how what amounts to "sovereign" power is exercised in the context of preemptive security. Moreover, recall from the previous chapter how the broader turn to anticipatory governance has been a direct response to the perceived failings of the traditional logic of state sovereignty vis-à-vis the problem of governing temporal contingency. Because the resulting emergence of anticipatory governmental strategies which move beyond the basic spatial logic of state sovereignty in an effort to tame time are the primary concern of this study, a traditionally limited understanding of the concept of sovereignty would be an inappropriate basis for any attempt to critically interrogate such strategies in the security context and beyond.

Thus, an understanding of "sovereignty" that takes a more Butlerian than Bodinian approach as its basis is more apposite, and subsequent use of the term "sovereign" or any of its derivatives should be understood in this more expansive way—as referring to any agent or actor that may be "delegated with the power to render unilateral decisions" in a particular context (Butler 2006, 56). To be sure, this certainly still includes the traditional sites of power such as the executive branches of state governments—and in fact much of the subsequent analysis will focus upon the behavior of the American executive and its preemptive prosecution of the War on Terror. The point, however, is that the understanding of sovereignty which underpins this primarily conceptual analysis *also* includes those myriad "petty" figures whom the exigencies of preemptive security often require to exercise their own form of sovereign power. Put differently, a core aim of part II of this book is to analyze and unpack the idea of preemptive security in the abstract, particularly as it relates to the organization and exercise of political power. As such, the arguments are developed in such a way that they should be understood to apply to *any* preemptive security decision—whether made by the president of the United States ordering a drone strike, a Frontex official refusing to admit a North African refugee into the European Union, an Indian counterterrorist intelligence officer ordering the preemptive detention of a suspected militant, or a Russian military commander ordering a preemptive strike against Chechen insurgents. Given that, following Butler, each of these security actors can be considered "sovereign" within the particular context of such decisions, a more expansive understanding of the term that recognizes this potential diversity is required.

The Globalization of Preemption

The preceding points about the generality of the analysis and the diversity of actors and agents involved in the practical implementation of preemptive security rationalities relate to a final consideration that must be addressed before proceeding. This concerns the question of the book's analytical scope and the potential (in)validity of the sort of general conceptual arguments it aspires to make. In this respect, two points must be addressed. The first relates to the *type* of analysis that will be conducted in the remaining chapters. On this point, I should reiterate that the arguments are concerned with critically theorizing the logic of preemptive security in the abstract, with a view to gleaning insights into the broader logic of anticipatory governance. As such, the aim is decidedly *not* to catalog particular examples thereof or conduct any in-depth empirical analysis of any such instances. Instead, I am primarily interested in expounding how preemptive security functions as a political rationality, and not in explicitly critiquing any specific practical manifestation thereof. As mentioned in the introduction, I draw upon multiple illustrative examples to empirically ground the core arguments; however, it should be emphasized once again that these arguments are primarily conceptual rather than empirical. The aim is thus to develop a critical account of the sort of politics that is logically presupposed and thus practically made possible by the adoption of a preemptive approach to (in)security governance, *irrespective* of the particular politico-geographical context in which this takes place. In other words, the arguments are framed in such a general way that they may be applied to *any* particular instance in which the logic of preemption informs the governance of (in)security. The reader is thus urged to bear the intentional generality of the analysis in mind when considering the claims advanced below, and to thus recall that the contributions I hope to make are primarily of a theoretical nature. Moreover, returning to the question of ontology, this framing is consistent with the sort of analysis to which "political rationalities" such as preemptive security can be subjected, as I attempt to interrogate the general pattern of operation through which preemptive approaches to (inter)national security problems are practically implemented. Put most simply, then, the aim is to theorize and critically interrogate the operational "regularities" that flow logically from preemption's conceptual constitution as a political rationality (Rose and Miller 1992, 178). As such, the conclusions reached should conceivably apply to all situations in which such a rationality forms the basis for practical policy.

This relates directly to the second point I wish to make, which concerns the geographic scope of the study and the effect this may have on the validity of such general conceptual analysis. In this respect, I recognize that this book's pretensions to conceptual generality are somewhat in tension with the empirical reality that the post-9/11 articulation of preemption upon which its analysis is primarily based has been largely manifested in the national security policies of Western states. Indeed, it is certainly the case that the contemporary articulation of preemption discussed earlier in this chapter refers largely to the strategies developed by the United States, United Kingdom, European Union, and other states of the West and the Global North in response to the threat of transnational terrorism. Moreover, the majority of the illustrative examples used to ground the conceptual arguments I develop in part II are also drawn from the practices of Western polities. It may therefore appear that the features I subsequently describe as generally applicable to logics of preemptive security in fact refer only to the parochial practices of Western states, suggesting that the conceptual generality of the claims is overstated at best and untenable at worst.

This is certainly a valid concern. However, there are two reasons that a focus upon the practices of Western polities is not incompatible with the sort of conceptual generality to which this book aspires. The first is that, simply put, the polities of the West offer far more examples of preemptive security in practice. Western states have been at the forefront of (re-)introducing preemption into the lexicon of (inter)national security, as first the United States under George W. Bush, and subsequently the United Kingdom under Tony Blair and the European Union in its Common Security and Defence Policy, adopted explicitly preemptive approaches to the governance of terrorism in the immediate post-9/11 era, which have been subsequently continued (de Goede 2008). As such, there are simply more practical instances of preemptive security to be found in the actions of Western states. It thus follows that any attempt to develop a more general theorization and conceptual interrogation of this paradigm of (in)security governance will draw more heavily upon examples derived from the practices of the West. The subsequent analysis should thus not be accused of unwarrantedly passing off the particularistic practices of Western polities as unproblematically representative of a general concept. Such accusations would miss the point, as it is my overarching interest in the general concept of preemption—specifically as applied to the governance of (in)security—that renders a certain West-centrism unavoidable. In other words, while the illustrative examples used to flesh out the conceptual arguments developed below will admittedly be drawn largely from the practices of Western state security apparatuses, this is a function of the empirical terrain itself, rather than of methodological imprecision, since it is in the polities of the industrialized West that preemptive security rationalities have been most comprehensively developed and implemented in the post-9/11 era, and which therefore provide the most productive basis for conceptualizing such rationalities in the abstract.

This is not to suggest, however, that the sorts of preemptive security rationalities with which I am concerned are *only* present in the West. Quite the contrary, as this turn toward anticipatory security has become a truly global phenomenon, particularly in recent years (Elmer and Opel 2008, 13). Crucially, however, when non-Western governments have adopted preemptive strategies and practices, these have tended to closely mirror those of the Western states mentioned above. This is the second reason that relying primarily upon the practices of Western states need not limit the study's potential for analytical generality, in that the anticipatory rationalities adopted by non-Western states have largely embodied the same core conceptual features as those prevailing in the West.

As an example, consider India, which has taken a number of steps toward institutionalizing a preemptive approach to the governance of terrorism over the past decade and a half (Kalhan et al. 2007, 141–72; Krishnan 2004). The 2002 Prevention of Terrorism Act (POTA)—which was passed in the aftermath of both the 9/11 attacks and the December 2001 attack on the Indian Parliament, and which lapsed in 2004—along with its successors, the 2005 Unlawful Activities Prevention Act (UAPA) and the 2008 National Investigation Agency Act (NIA), all contain provisions designed to enhance the state's capacity to anticipatorily govern terrorism. For instance, all three statutes entrenched the state's prerogative to exercise "preventive detention" against suspected terrorists—for up to 180 days in the case of the UAPA (Mate and Naseemullah 2010, 263)—while remaining "silent on the evidence required" to activate this prerogative (Krishnan 2004, 283). Such a combination of a preventive posture and a limited evidentiary threshold for anticipatory action suggests a broadly precautionary ethos that is clearly reminiscent of the post-9/11 formulation of preemption described earlier in this chapter, and which closely resembles the practices of Western states. This is in fact no coincidence, as POTA was explicitly modelled after the USA PATRIOT Act and the British Anti-terrorism, Crime and Security Act—both of which served as the legislative foundations for preemptive security rationalities in their respective states—with Indian lawmakers even enlisting American officials to help formulate the final draft of the law (Sasikumar 2010, 629). India's nascent preemptive security regime thus quite conspicuously resembles that which has emerged in the West.

A similar observation can be made with respect to another important non-Western state, Russia. Indeed, since Vladimir Putin's ascent to the presidency at the turn of the millennium, the Russian government and security apparatus have demonstrated an affinity for preemptive tactics and strategies with respect to the governance of terrorism—particularly vis-à-vis Chechnya (Elmer and Opel 2008, 13; Guertner 2007; Westphal 2003). Importantly, this affinity became more pronounced following the Bush administration's overt declaration of a preemptive approach to the War on Terror in 2001, as "preemption quickly found its way into Russian declaratory policies encompassing both domestic and international threats" shortly thereafter (Guertner 2007). This period also coincided with an escalation of the Chechen insurgency, the Russian response to which also evinced a preemptive turn reminiscent of that which was taken by the United States under Bush. For instance, in the wake of the 2004 Beslan hostage crisis, then Chief of the General Staff Yuri Baluevsky explicitly echoed the language of the Bush administration and "declared the right to launch preemptive strikes against terrorist bases worldwide," while Russian security officials have continually "invoked preemption to justify future interventions against Chechen separatists" (Guertner 2007). Moreover, after regaining the presidency, Putin himself reaffirmed his commitment to a preemptive approach to the governance of terrorism, remarking at an October 2012 press conference that Russia will continue to "pursue a preemptive tactic" against terrorists, and emphasizing his government's successes in this regard by enumerating the number of "militants" that had been "neutralized" through preemptive targeted killing in recent months (Bridge 2012).

Russia and India's apparent embrace of the same sorts of preemptive security strategies developed by Western states further emphasizes that the worldwide proliferation of such strategies is a key dynamic of the contemporary global security climate. However—and more important for present purposes—it also suggests the analytical validity of using the practices of Western states to empirically ground a set of conceptual arguments relating to the logic of preemptive security in the abstract. Indeed, the Indian and Russian examples suggest that the preemptive security rationalities adopted by non-Western states share the same broad conceptual contours as those that increasingly underpin the practice of (in)security governance in the West. Accordingly, a critical conceptual account of preemptive security as manifested in the post-9/11 era can be legitimately developed with the aid of illustrative examples drawn primarily from the practices of Western state security apparatuses.

CONCLUSION

Developing such an account is the task of the remainder of this book, and the two chapters of part I have attempted to lay the foundation for such an undertaking. Toward this end, this chapter in particularly has elaborated upon what preemptive security should be understood to mean in contemporary context, and has established some basic analytical parameters within which it can be critically interrogated in conceptual terms. Having thus considered what preemptive security is and how it can be examined conceptually, the book's core arguments relating to how logics of anticipatory governance affect the organization and exercise of political power can now be more fully developed through a detailed theorization of the politics of preemptive security. The next chapter begins this investigation by considering in further depth how the post-9/11 formulation of preemption operates as a political rationality in the security realm.

NOTES

1. I will hereafter use this term when referring to the set of ideas that inform the anticipatory approach to (in)security governance with which I am concerned. Since the arguments are developed primarily from a general theoretical perspective, such an all-encompassing term is required for the analysis. Moreover, using the term in this way is consistent with existing terminological practice in the literature (see Anderson 2010a, 790; Massumi 2007, 9).

2. Again, given the conceptually general focus of the subsequent analysis, I use this term as something of a catchall to refer to the array of security strategies and practices that are united by their basis in the logic of preemption. This is also consistent with prevailing terminological practice in the literature (see, for instance, de Goede 2011, 2008; Sullivan and Hayes 2010).

3. Before proceeding, I should emphasize that the aim of this discussion is to understand how preemption is conceptualized *as an idea* in the discourse of international law. A potential obstacle to this aim is that treatments of preemption in international law formulate abstract *conceptual* articulations of preemption via *normative* arguments relating to the legitimacy of anticipatory self-defense. In other words, the abstract meaning of preemption as an idea is embedded in a set of claims about the type of anticipatory action that is justifiable under natural, customary, or positive law. It must therefore be made clear that in what follows I am not concerned with evaluating the cogency of the normative arguments in favor of a certain understanding of preemption. Rather, I seek merely to use some of the more notable of these normative statements as the basis for teasing out the abstract conceptual understanding of preemption embedded in the discourse of international law, so as to subsequently compare it to the conceptualization articulated in the post-9/11 War on Terror.

4. Chapter VII, Article 51, affirms the right to self-defense, but does not state whether this right extends to preemptive action. However, its specification that this right may be invoked only "if an armed attack occurs" would seem to indicate that, in fact, it does not. On this point, see Arend 2003, 91–93.

5. The question of political "exceptionalism" in general, and its relationship to preemptive security in particular, is explored at length in chapters 5 and 6.

6. According to a 2014 report by Human Rights Watch, "Nearly 50 percent of the federal counter-terrorism convictions since September 11, 2001, resulted from informant-based cases," while "almost 30 percent were sting operations in which the informant played an active role in the underlying plot" (Human Rights Watch 2014, 21).

7. This exemplifies the phenomenon of "counter law" serving as an enabler of anticipatory governance mechanisms (Ericson 2008), which will be discussed at length in chapter 5.

8. A more specific elaboration of precisely the sorts of actors that this refers to will be undertaken shortly.

9. However, in a very interesting modification of Butler's framework, Roxanne Lynn Doty (2007, 2009) has extended the concept of the "petty sovereign" to include agents acting entirely outside the formal governmental apparatus. Using the case of so-called Minutemen militia groups that have formed along the US-Mexico border for the purpose of tracking and deterring irregular migrants, Doty demonstrates how private citizens have enacted themselves as what amount to sovereign decision-makers in this particular context, arrogating to themselves powers conventionally limited to the state by making decisions about inclusion and exclusion in the American polity and occasionally using violent means to enforce these decisions.

Part Two

It is necessary not only to pay attention to immediate crises, but to foresee those that will come and to make every effort to prevent them.

—Machiavelli, *The Prince*, chapter III

Nothing is so wretched or foolish as to anticipate misfortunes. What madness is it to be expecting evil before it comes.

—Seneca, *Epistulae Morales ad Lucilium*, XCVIII

I have the imagination of disaster—and see life as ferocious and sinister.

—Henry James, letter to A. C. Benson

Chapter Four

Anticipatory Political Timescapes

The preceding chapter offered a comparative conceptual analysis of the notion of preemption as applied to the realm of (inter)national security, where it was argued that the post-9/11 articulation is qualitatively different from its antecedents in the realm of international law, primarily due to its explicit embrace of what amounts of a precautionary logic that takes radical uncertainty as the basis for, rather than an impediment to, anticipatory action. Having thus considered what preemption "means" in the current global security context, the focus of this chapter will be exploring how preemption "works" as a political rationality applied to the realm of (in)security governance. In other words—and to once again invoke Rose and Miller—I am concerned with theorizing the "regularities" of this formulation of preemption as a mode of governance, with a view to developing a detailed conceptual account of precisely what logics are at work in its general pattern of operation. This first chapter of part II adopts perhaps the most abstract perspective of the three that comprise it, serving as something akin to background for the more narrowly targeted analyses of chapters 5 and 6. This does not render the arguments developed here any less important to the overall goals of this book, however. On the contrary, a wider understanding of how the contemporary conception of preemption functions as a political rationality is necessary for understanding how anticipatory governance strategies based thereupon modify both the organization and exercise of political power and the experience of political subjectivity—issues that will be examined respectively in the next two chapters.

The remainder of this chapter is thus concerned with conceptually exploring the operational logics of preemption as a political rationality. This discussion will be centered around two particular aspects thereof, the exploration of which will be the respective focus of the two subsequent sections. The first reflects this study's emphasis on foregrounding temporality and futurity, in that it explores how a politics of preemption operates by effectively manipulating experiences of time itself. In this respect, it is argued that preemptive logics of governance enact a temporality in which the future itself is "made present" in such a way that its contingent potentialities serve as the primary causal agents of present political action. Such a temporality implies a significant rearticulation of our relation to the future, in that all future potentialities—no matter their likelihood of occurrence or temporal distance from the present—must be considered potentially imminent. This is because only then can the future's radically uncertain contingencies trigger the sorts of affective responses that are required to render the merely potential practically actionable in the present. The key claim of the first section is thus that a politics of preemption works by manipulating our subjective relation to the future in particular, and time more generally.

By focusing on the theme of temporality and touching on the question of affect, this discussion segues into the second section, which begins by discussing how the temporal manipulations outlined in the first shape the praxis of politics. Here it is argued that preemption's expansion of the horizon of imminence serves to compress the timescape of political decision-making in a manifestly anti-democratic fashion. This is because a seemingly perpetual imminence of catastrophe demands not only that anticipatory action be taken but also that it be taken as soon as possible, thus militating against the deliberative processes that are central to democratic politics but invariably take time to carry out. It is then argued that this compression of the political timescape elevates the importance of the affective plane in political practice. This is because demands for immediate action lead decisions to be made primarily on the basis of "gut feelings," since there is inadequate time for the diligent collection and deliberative analysis of empirically verifiable facts and evidence. In light of these arguments, I conclude that the adoption of a preemptive approach to (in)security governance has significant implications for the way political power is organized and exercised—a point that is then taken up at much greater length in chapter 5.

TEMPORAL MANIPULATIONS

The Presence of the Future

Melinda Cooper concisely explains the core premise of the logic of preemption as articulated in the context of contemporary (in)security governance: "it exhorts us to respond to what we suspect without being able to discern; to prepare for the emergent long before we can predict how and when it will be actualized; to counter the unknowable before it is even realized" (2006, 120). As discussed in the preceding chapter, perhaps the defining characteristic of a political rationality based on this logic is its highly temporal orientation, in that its primary concern is the (re)assertion of control over the vicissitudes of time by actively controlling its unfolding in a manner that diminishes the potentiality of catastrophe associated with a condition of radical uncertainty. In this sense, adopting a preemptive approach to societal governance can be understood most generally as an attempt to "act on time" itself (Massumi 2005a, 5). More specifically, preemption is characterized by a constitutive futurity. Indeed, as is apparent from both Cooper's description and the account developed in the preceding chapter, the logic of preemption is fundamentally oriented around the question of the open future and its concomitant unknowability, since it aims to confer upon human agents a capacity to govern the future's potential catastrophes through appropriate anticipatory interventions in the present. Thus, rather than simply acting on time in general, the logic of preemption as articulated in the contemporary (inter)national security context might be better understood as concerned with acting on the future in particular, in that "possible futures are brought into the present to be acted upon" (Adey, Anderson, and Graham 2015, 10). As Leese puts it, "[I]n post-9/11 security regimes, the efforts of policymakers to capture the future and fold it back into the present in order to make it actionable have reached new heights" (2014, 497).

This centrality of the future to the logic of preemption is crucial to recognize when thinking through how the latter functions as a political rationality. This is because such an emphasis on the future has the effect of significantly reconfiguring the timescape within which political decisions are made. It does so, in particular, by enacting a condition in which the unknowable future is *prioritized*—both in terms of the creation of knowledge on the basis of which decisions are taken and in terms of the location of the desired outcomes of those decisions—while the knowable present is instrumentally *marginalized* as the mere location of the requisite interventions.[1] In other words, under a preemptive political framework, decisions are made primarily with reference to a potential (and thus imaginary) future that may or may not come to pass, rather than on the basis of present (and thus empirically verifiable) conditions, and the primary goal is to eliminate that future as a temporal potentiality rather than to chart a particular normative course. The logic of preemption shifts the temporal orientation of political decision-making from what Brian Massumi terms a "past-present axis"—concerned with "trends whose arc will continue more or less predictably into the future" and thus permit the active pursuit of a particular developmental path—to a "present-future axis"—which "wobbl[es] with uncertainty" and is thus concerned less with charting a particular course than with fending off potentially catastrophic irruptions that may or may not occur as the future unfolds (Massumi 2005a, 4). In this sense, the crucial temporal relationship of a preemptive politics is between the present and future; yet this relationship is (re)articulated in such a way that the future is no longer construed as the somewhat benign *tabula rasa* of Enlightenment thought, whose openness renders it amenable to the creative energy of the human present in a manner enabling continuous positive progress (Koselleck 1985). Rather, within the timescape of preemption, the future looms large and darkly over the lived present, with its radical contingency and catastrophic potentiality shaping key decisions made therein—decisions that are directed primarily toward preventing, rather than realizing, the possible. In short, the logic of preemption rewrites the political timescape in such a way that the *potential* future is a primary causal agent in the *actual* present, rather than vice versa. It is in this respect that a politics of preemption can be understood to operate through the manipulation of our experience of time itself.

These points become clearer when considering the mechanisms through which preemptive political rationalities function in greater detail. Massumi's (2005a, 2005b, 2007) readings of such questions are particularly useful, as he emphasizes how the logic of preemption is premised upon mobilizing merely potential events located in the depths of the unknowable future to produce actual consequences in the lived present. This phenomenon is perhaps best described by Ben Anderson as the "presence of the future" (Anderson 2010a, 783), in that preemption "brings the future into the present . . . mak[ing] present the future consequences of an eventuality that may or may not occur" (Massumi 2005a, 8). Indeed, under the logic of preemption, a future irruption that may never actually take place—and, regardless, has no ontological presence outside the imagination of those considering it—enacts very real consequences in the present through the interventions that follow from an anticipatory act undertaken to preclude its coming to pass. The actual future occurrence of the catastrophic irruption is ultimately irrelevant, since the preemptive decision grants it causal and/or constitutive purchase in the present by casting it as the basis for anticipatory action that can only take place in the present, and which will prevent its emergence in any case. Under the logic of preemption, therefore, the potentially catastrophic "event is an eventuality that may or may not occur, but *does nevertheless in effect*" (Massumi 2005a, 9, emphasis added). In other words, while the catastrophe remains always *virtual* and located in the unknowable future, the logic of preemption ensures that it nevertheless enacts very *actual* effects in the present. Thus, even though it "has not happened and may never happen," it is as if it has already occurred—it occurs "in effect" (Massumi 2007). The imagined catastrophic event can be understood as both present and absent, in that it exists only in the virtual realm of the contingently possible, yet it brings into being very tangible material consequences in the lived present of the actual.

It becomes possible to see how the political logic of preemption operates through the manipulation of time, in that it rewrites relationships between present and future in a way that destabilizes dominant articulations of time in the context of political decision-making. In this respect, the "presence of the future" subverts the conventional narrative of time as a linear unfolding—whereby the "arrow of time" flows from the past through the present and into the future, with causal relationships following the same direction—in that the always-to-come future is granted what amounts to causal power vis-à-vis the already-here present. The conventional logic of cause-effect relationships congruent with such a linear temporality is challenged, as an indefinite future event engenders tangible effects in the present without yet having taken place. Preemption thus creates something of a "time-slip," whereby the future in general—and any potentially catastrophic irruptions deemed to inhabit it in particular—recursively "slips" outside of the linear flow of time and reenters it in the present through the precipitation of anticipatory interventions therein. Put another way, potential future events become "transtemporal" under a preemptive political rationality (Massumi 2005b, 36), taking on a dualistic ontology somewhat reminiscent of Heisenberg's electron in which they possess an effective presence in *both* the potential future *and* the actual present. By constructing such a timescape, in which the future is able to directly generate material consequences in the present, the logic of preemption inscribes an alternative temporality in which cause-effect relationships deviate from the strictures of temporal linearity by, in a sense, reversing the causal arrow such that it doubles back against the proverbial arrow of time.

To clarify these points through an illustrative example from the global security realm, briefly consider that most prototypical instance of post-9/11 preemption, the American decision for war in Iraq. In this case, key decision-makers conjecturally imagined a future in which Saddam Hussein's Iraq *had* covertly achieved weapons of mass destruction (WMD) capability, which became verified fact only when the "smoking gun" came in the form of a "mushroom cloud" (Ehrenberg et al. 2010, 552).[2] The decision was made to intervene preemptively in an attempt to assert control over the unfolding of the future in such a way that this potentiality was actively eliminated from the realm of possible futures. This process gave the potential future event of a hostile Iraq wreaking global nuclear havoc causal purchase in the present, and subsequently generated profound tangible consequences in the form of a large-scale military incursion that radically altered the geopolitical realities of that present. The American decision for war thus shows how a preemptive politics functions through a reconstruction of the timescape of the political, in which the temporal-causal relationship between present and future is reformulated in a significant way. And while the Iraq invasion is perhaps the most well-known application of the post-9/11 formulation of preemption, the logic of anticipatory decision-making that it illustrates has by no means been limited to the somewhat idiosyncratic "Bush Doctrine" of interstate war. Indeed, similar rationalities have been widely employed across the broader War on Terror (Amoore and de Goede 2008), thus enacting this same process of temporal manipulation in a variety of anticipatory governance contexts—from the practice of indefinitely detaining individuals merely suspected of terrorist action or sympathy (Ericson 2008; Mutimer 2007), to the preemptive freezing of mobile capital suspected of terrorist links (de Goede 2011, 2012), to the extrajudicial killing of terrorist figures deemed threatening to a particular state (Grayson 2012; Kessler and Werner 2008).

It should also be noted that these temporal manipulations are equally present in the anticipatory decisions made by the "petty sovereigns" often tasked with carrying out preemptive security policies. Accordingly, the conceptual points illustrated by the Iraq case can be understood as applicable to *any* circumstance in which the logic of preemption is adopted as a framework for the governance of (in)security. Moreover—and returning to a broader theme of the book—this discussion also illustrates the importance of adopting a temporal lens to properly understand how the logic of preemption operates as a political rationality. Indeed, only by foregrounding temporality do we see that a crucial dimension of preemption's operation consists of reconstituting the political timescape in such a way that the potential future acts as a primary determinant of the actual present. This represents a significant departure from prevailing linear understandings of time and associated relations of agency and causality that implicitly underpin conventional logics of political decision-making.

While perhaps interesting on their own terms, these abstract considerations must be further unpacked prior to proceeding with a more targeted analysis of preemptive security as it is enacted in the global War on Terror. Indeed, now that we have seen how preemption operates through the manipulation of experience of time, the question becomes: How are these processes of temporal manipulation mediated through the behavior(s) of political decision-makers? In other words, it is not enough to merely describe the temporal manipulations that are logically inherent in a politics of preemption; it is also necessary to examine how these temporal manipulations work to legitimate anticipatory action among public authorities and citizenries. This requires a more detailed consideration of the way humans as political subjects relate to the future, since the recursive "time-slips" through which preemption becomes politically operationalized are ultimately made possible by the way we apprehend and engage with the future from the vantage point of the lived present.

Imminence and the Passions

The manner in which a politics of preemption relies upon a particular way of apprehending the future can be usefully fleshed out through an engagement with Philip Fisher's typology of temporal spaces developed in his 2002 book *The Vehement Passions*. In a chapter titled simply "Time," Fisher offers an account of the way "the passions"—meaning the affective resonances that color and characterize the human experience—are crucial to mediating our understanding of, and relation to, time in general and the future in particular (Fisher 2002, 78). Of particular relevance here is Fisher's distinction between what he terms the "abstract future" and the "imminent future," and their respective relationships to the present. Fisher describes the abstract future as that which is so temporally distant that it can only be apprehended through the exercise of the imagination, since it is too far removed from our existence in the present for us to possess any coherent knowledge or understanding of what it holds (79). As a consequence, the passions are rarely activated by consideration of the abstract future. Indeed, this is the abstract future's defining feature for Fisher, since the epistemic ambiguity stemming from its temporal distance means that "we cannot think of ourselves as acting reasonably about" it; as such, no sense of acute urgency regarding its potentialities—which manifest through such affects as fear, hope, or desire—is felt, and no subsequent impetus to act is generated (80). Put more simply, Fisher's point is that the temporally distant abstract future is insufficiently intelligible to human subjects to generate the types of affective responses that he claims are required to induce us to take action in specific relation to it.

Conversely, the imminent future refers to the temporally proximate, about-to-be-realized moment—reflecting, for instance, the sort of all-but-certain imminent event with which the international legal understanding of preemption is concerned. In contrast to the abstract future, Fisher contends that the imminent future's temporal proximity grants it an adequate degree of intelligibility to make "rational" action based upon our apprehension of it seem possible, which activates the sorts of affective responses that generate the impetus for present interventions in relation to it (Fisher 2002, 80–81). In other words, the imminent future is demarcated from the abstract future at precisely that point where the passions are triggered by a specific apprehension of what is to come, such that we feel able to act in relation to that to-come. As Fisher puts it, "The line between the abstract future and the imminent future is drawn by the moment at which fear or yearning, hope or dread, engages" (78).

Fisher articulates this distinction between the abstract and imminent future—and the role of the passions in constituting this difference—most clearly with specific reference to the effect of fear: "We feel fear in the face of what is just about to happen. . . . [A] possible harm located in the indefinite [abstract] future cannot be the subject of fear (or hope), but it is precisely fear that tells us where the indefinite future leaves off and the imminent, or definite, future begins" (2002, 72, 78).[3] Of course, fear is by no means the only affective response capable of demarcating between these two futures and thus triggering the impetus for present action. A more amorphous affect of general anxiety wrought by the uncertainty of the impending future could, for instance, also act as the trigger, as could an affect of desire relating to the urge to bring into being one particular iteration of the future in lieu of another. Indeed, read more broadly, Fisher's framework suggests that a combination of affective responses, unique to each circumstance, coalesce to generate the impetus to act in response to what is to come, thus demarcating the imminent future and rendering it anticipatorily actionable.

The key point for present purposes, however, is that with respect to the imminent future, our perception of the potential becomes presently actionable. In other words, even though we can never be *entirely* certain as to the precise details of what is about to occur, Fisher's schema suggests that the imminent future's temporal proximity grants it an intelligibility that makes reasoned action framed in relation to it appear possible, which in turn triggers the sorts of affective responses that can precipitate action of some sort (2002, 78, 86). And because the triggering of the passions is deemed a necessary condition for action in the present, the *abstract* future—which is accessible only through the exercise of a speculative imagination—should *not* be considered actionable in the present, since it is *explicitly defined* by the lack of any significant affective response to its potentialities. Paraphrasing Hume, Fisher summarizes these arguments by asserting that "the passions are felt only for things contiguous or proximate in time"; as such, the "neighbourhood of time surrounding the present within which the passions operate"—specifically, the immediately imminent future—is the only temporal space from which potential events can generate the degree of affective resonance required to trigger some sort of human action (76).

To bring all of this back to the politics of preemption, Fisher's framework suggests that a decision to act on the basis of that which is yet to come—that is, a preemptive or anticipatory decision—can only be made with respect to the imminent future, since only then can humans generate an intelligible enough picture to make such action appear reasonable and/or possible and thus activate the affective responses that precipitate a decision to act. Fisher's affectively mediated account of our relation to the future provides a psycho-philosophical grounding for the more "traditional" articulation of preemption associated with the discourse of international law discussed in the previous chapter, as the latter is also premised upon a short temporal horizon and a high degree of confidence with respect to what is about to take place. Indeed, the type of action presupposed by such understandings of preemption consists of taking anticipatory measures in relation to a potentiality located in Fisher's imminent future. Conversely, the sort of potentiality that provides the impetus for anticipatory action under the post-9/11 articulation of preemption inhabits Fisher's *abstract future*, since the condition of radical uncertainty that this formulation is premised upon confronting implies that the potentiality in question is confined to highly speculative imaginings about what may come to pass at some indefinite point. Crucially, Fisher's framework suggests that this should render such a potentiality effectively inactionable in the present, since it ostensibly cannot be apprehended with adequate intelligibility to generate the sort of affective response that would precipitate action. According to Fisher, in other words, the post-9/11 politics of preemption should be impossible; yet we have seen that it takes precisely these sorts of apparently inactionable potentialities as the very basis for action.

Readers may at this point be questioning why Fisher's ideas have been discussed at all, since they seem manifestly incompatible with—perhaps even effectively ruling out—precisely the sort of preemptive political rationalities that are the focus of this book. Yet reading the politics of preemption through Fisher's framework in fact further illustrates the key claim that preemption functions as a political rationality through the manipulation of perceptions of time. In particular, it shows how a politics of preemption requires a radical modification to the way human subjects relate to the future. To elaborate, if we take seriously Fisher's claim that affective responses are crucial to precipitating any type of human action—of which anticipatory governance decisions are certainly an example—and that the imminent future is the only portion of the to-come whose potentialities can adequately activate the passions in this way, then it follows that a politics of preemption must alter our relation of the future to the extent that the distinction between abstract and imminent future collapses and the passions can be activated with respect to *any and all* potentialities, irrespective of their temporal proximity to the lived present. In other words, the logic of preemption rearticulates our subjective relation to the future in such a way that the "imminent" portion thereof is effectively extended indefinitely. This has the effect of moving the radically uncertain potentialities of the abstract future—whose epistemic fuzziness Fisher claims disqualifies them from activating the passions and thus precipitating action in the present—into an indefinitely extended imminent future, where they can now become fuel for the affective responses of decision-makers and thus be capable of inducing anticipatory political interventions. In this sense, under the contemporary logic of preemption, the far-off specter of potential catastrophe—which still remains inherently imaginary and entirely virtual—is no longer a mere abstract potentiality too far removed from the present to be rationally actionable; rather, it is rearticulated as an always imminent threat which must be countered anticipatorily. The centrality of this particular epistemic move to the post-9/11 formulation of preemption is quite clearly demonstrated in the wording of the 2002 US *National Security Strategy*, as this text's outline of a preemptive security rationality famously makes an explicit call to "adapt the concept of imminent threat" to include those merely potential threats that may emerge at some indefinite future point (see White House 2002; Ehrenberg et al. 2010, 83). Recalling Massumi's claims discussed above, this can be viewed as another instance of the logic of preemption creating a "time-slip" through which temporality is manipulated to make anticipatory interventions possible in the face of radical uncertainty about a potentially catastrophic future.

The takeaway point, therefore, is that the logic of preemption alters our relation to the future to the extent that—at least with respect to such potential catastrophes as the "next terrorist attack"—*all* potentialities, regardless of their degree of practical intelligibility or temporal proximity, are effectively treated as imminent and as legitimate stimuli for anticipatory action. Here we again see echoes of the precautionary principle, which is premised upon taking anticipatory action under the assumption that all imaginable potentialities are possible (Ewald 2002; Sunstein 2007). But again, this does not imply that all such potentialities *will* precipitate anticipatory interventions, since decisions must still be made as to precisely which possible futures *should* be acted upon and how—decisions that are highly political and will inevitably reflect broader practices of threat construction informed by prevailing societal prejudices and discourses of othering (Campbell 1998). This highlights an important distinction between the conceptual politics (or logic) of preemption—which is the primary concern of this book—and the empirical politics of preemption—which concern the substantive character of such decisions, and have been explored extensively in critical security studies literature. My primary concern is merely to show conceptually how preemptive political rationalities create the *conditions for* such potentially problematic decisions by reframing all potentialities as effectively imminent and thus rendering them capable of triggering affective responses that provoke an impetus to act.

To sum up, under the post-9/11 formulation of preemption, Fisher's distinction between the imminent and abstract future collapses, as the former is effectively extended indefinitely into the latter. This enables all imaginable possibilities to become potential triggers for the affective responses necessary to precipitate action, thus rendering the radically uncertain and temporally distant future anticipatorily governable in the present—which, of course, is the ultimate normative aim of a politics of preemption. Importantly, these points correspond directly to the arguments of the previous chapter, where it was shown that the contemporary articulation of preemption in the context of (inter)national security extends the temporal horizon within which a potential event can be considered so imminent as to require anticipatory action. Indeed, the temporal sleight-of-hand in which catastrophic potentialities are transported from the abstract future of mere imagination to the imminent future of urgency and affective resonance is precisely what enables the logic of preemption to "make the future present" and transform radical uncertainty from a *barrier to* action—as under both the conventional articulation of preemption and Fisher's framework—into a politically legitimate *basis for* action. Mapping the logic of preemption on to Fisher's affectively oriented temporal taxonomy further highlights how the manipulation of time in general, and the reframing of our relation to the future in particular, are crucial to how the logic of preemption functions as a political rationality.

The discussion so far has drawn attention to how preemption's operation as a political rationality is intimately bound up with the way temporality is understood and experienced, and the importance of the affective realm to such questions. Yet there is more to how these issues operate within the politics of preemption than has thus far been shown. With respect to our experience of time, despite outlining how a preemptive politics functions through a series of temporal manipulations that alter our subjective relation to the future, the above discussion does not address how this change in the phenomenological timescape also alters the timescape of political praxis. In other words, we have not yet considered how the logic of preemption's temporal manipulations also extends to the deliberative environment in which decisions are actually made. This elides the extent to which a preemptive politics compresses the political timescape in an ultimately undemocratic way. With respect to the importance of affect, the above discussion highlights the role of the affective in a highly general way that merely affirms its importance to all forms of human decision-making, anticipatory or otherwise. Such a limited discussion is incomplete in the present context, since—for reasons directly related to the aforementioned modifications of the political timescape—the importance of the affective is especially acute in the case of anticipatory political decision-making. To continue this exploration of how preemption functions as a political rationality, both of these points must be unpacked further, since the anti-democratic compression of the political timescape that results from preemption's temporal manipulations and the concomitant importance of affect in the decision-making process are crucial to understanding how a politics of anticipatory governance operates in practice, both in the security realm and beyond.

COMPRESSED TEMPORALITIES

By modifying our relation to the future in such a way that *all* future potentialities are construed as imminent and thus capable of precipitating anticipatory decisions in the present, preemptive governance rationalities have the effect of compressing the political timescape in which these decisions are made, with ultimately anti-democratic consequences. Indeed, because preemption assumes that the potentialities against which it seeks to act could emerge at any time, it follows that minimal time can be devoted to such tasks as evidence collection and democratic deliberation in the context of anticipatory decision-making, since every moment spent deciding whether and how to act is a moment in which the potentiality whose avoidance is sought could come to pass. This dynamic of temporal compression can be considered a second-order temporal manipulation that follows from the first order manipulations in which the "presence of the future" is established through the indefinite extension of the imminent future, and it poses particular problems for the liberal democratic polities that have been in the vanguard of adopting anticipatory governance strategies premised upon some variation of the post-9/11 conceptualization of preemption.

To elaborate on these points, the practical politics of preemption most basically involve decisions about anticipatory action against a particular future threat; however, the temporal context in which these decisions are made—where the potentialities it seeks to preclude are construed as perpetually imminent—implies that the imperative to act in this way is driven by a sense of immediacy (see Adey, Anderson, and Graham 2015, 11; Cooper 2015). In other words, the logic of preemption dictates that we must act in the present to head off a potential future catastrophe, and we must *act now*, since the potential imminence of the imagined catastrophe in question—about which we can know relatively little due to the condition of radical uncertainty being faced—means that it could emerge at any moment. In the security context in particular, this compression of the political timescape has been recognized in official documents that articulate a preemptive strategy, and the statements of policymakers defending such an approach. For instance, the 2002 US *National Security Strategy*—arguably the foundational document of the post-9/11 notion of preemptive security—spoke of "the risk of inaction . . . even if uncertainty remains as to the time and place of the enemy's attack" (White House 2002), while we have already seen how former Vice President Cheney asserted that "the risks of inaction are far greater than the risk of action" in the case of the putative threat from Saddam Hussein's Iraq. Both pronouncements imply that a preemptive approach to (in)security governance demands the prioritization of immediate action over extended deliberation, since the radical uncertainty of the future it seeks to govern means that all delays merely increase the likelihood that the threat in question will catastrophically materialize. Cheney's words, in particular, were spoken in the context of arguing against giving UN weapons inspectors more time to collect verifiable empirical evidence regarding Iraq's alleged WMD program, since he argued that doing so "would increase the military risk posed by the Iraqi regime" (Daase and Kessler 2007, 426).

This implies that the collection and evaluation of information, and subsequent public deliberation about the appropriate course of action, must be limited under a politics of preemption. Indeed, there is *no time* for either, since every moment spent discussing the proper response to an always potentially imminent catastrophe is a moment in which that catastrophe might occur. A preemptive politics is thus characterized by "a collapse in time, contemplative time to be exact," such that a priority is placed upon the taking of (anticipatory) action as such, rather than ensuring that such action rests upon a sound epistemic foundation established through careful verification and deliberation (Elmer and Opel 2008, 14). The result is that, to use another of Massumi's apt turns of phrase, an anticipatory political decision "strikes like lightning" (2005a, 5), since it is undertaken in a compressed timescape that is likely to preclude any serious deliberative activity.[4]

This represents another way in which a preemptive political rationality differs markedly from other logics of political decision-making, and, more important, it suggests some potentially significant tensions between a politics of preemption and the constitutive norms of liberal democracy. In this regard, as Sheldon Wolin pithily reminds us, politics—particularly of the democratic sort—inevitably "takes time" (Wolin 1997, 2; see also Manning 2004); yet the logic of preemption's constitutive impetus to act now rather than later militates against this truism, such that "the elixir of speed" (Falk 2010, 255) becomes more important than any commitment to empirically and/or democratically oriented deliberation about appropriate action. This significantly compromises the democratic integrity of political decision-making in that the imperative to "act now" runs explicitly counter to the agonic deliberative processes associated with functional democratic governance, reliant as these are upon the extensive consultation of, and compromise between, a variety of actors and stakeholders. As Wolin argues, in "societies with pretensions to democracy," political action must "be preceded by deliberation, and deliberation, as its 'deliberate' part suggests, *takes time* because, typically, it occurs in a setting of competing or conflicting but legitimate considerations" (1997, emphasis added). Seyla Benhabib makes a similar point in a discussion of contemporary liberal democracy, asserting that legitimate political action "in complex modern democratic societies must be thought to result from the free and unconstrained public deliberation about matters of common concern"—a process that can only take place over an at least somewhat open-ended period, since the strict limits of a compressed political timescape would militate directly against the "unconstrained" character of such deliberation (Benhabib 1994, 26). Wolin thus concludes—with an implicit allusion to ancient republicanism—that a truly democratic politics "requires an element of leisure . . . in the sense, say, of a leisurely pace," since its constitutive deliberative elements cannot be legitimately enacted otherwise. Such considerations are also relevant to the legal rather than procedurally political elements of liberal democracy. As Opitz and Tellman put it, a liberal juridical order is "time-consuming," since "each voice has to be heard [and] evidence thoroughly tested" (2015, 121). Echoing Wolin, they assert that "taking time thus becomes a precondition for accurate legality" in a liberal democratic system.[5]

The temporal compression associated with a politics of preemption works to evacuate the polity in which it is being implemented of its democratic character, in that the type of "lightning decision" that it demands is fundamentally incompatible with the often languid processes and procedures of legitimate democratic decision-making, both political and legal. Moreover, not only does preemption compress the time for decisions to be made but, in an additional affront to the open deliberation of democratic politics, it also limits the terms of the debate regarding potential action from the outset, since the more basic question of whether to act at all is itself effectively predecided in the affirmative. This is again because preemption's reformulation of our relation to the future—such that all potentially catastrophic contingencies are construed as effectively imminent—demands that anticipatory action *must* be taken, regardless of prevailing uncertainty as to the specific character or likelihood of the threat(s) in question (Kessler and Daase 2008, 225–26).

Under such circumstances, therefore, "there is little need for public deliberation and debate" at all, since the perpetual potential imminence of catastrophe implies that there is no alternative but to act, and act now (Elmer and Opel 2006, 479).

These conceptual points can be illustrated by the character of American political debate and decision-making in the aftermath of the 9/11 attacks and in the lead-up to the archetypically preemptive invasion of Iraq. For instance, in a statement before the Senate arguing against the adoption of the controversial USA PATRIOT Act—whose provisions provided the backbone of the emerging preemptive security regime—Wisconsin Democratic Senator Russ Feingold raised concerns that highlight precisely these issues of anti-democratic temporal compression. Feingold cautioned against the extremely rushed passage of such sweeping legislation, expressing concern that "the pressure to move on this bill *quickly, without deliberation or debate*, has been relentless," and arguing that "the preemptive ethos in the White House created a 'demand for haste' that was 'inappropriate'" for a liberal democratic polity (quoted in Ehrenberg et al., 2010, 528, 531, emphasis added). Feingold's statements highlight not only how the temporal compression inherent to a politics of preemption affected the American response to the 9/11 attacks but also how this resulted in the diminished democratic character of the debate surrounding this response. A similar point could be made with respect to the (inter)national debate(s) that took place in the prelude to the Iraq War, as evidenced by Vice President Cheney's above-quoted assertion, which assumes that the question of whether to act at all is pre-given in the affirmative while any deliberative delay in action would prove manifestly dangerous (see Daase and Kessler 2007; Kessler and Daase 2008). The key point, however, is that the anti-democratic depoliticization evidenced in these instances is as much a structural result of the logic of preemption itself as it is a particular effect of the idiosyncratic policy decision(s) and local political environment of any specific case. As such, this should be considered a particularly important effect of adopting anticipatory approaches to governance problems of any kind.

These counter-democratic tendencies also have potentially profound implications for the individual human subjects against whom anticipatory interventions are deployed. On this point, the 2005 fatal shooting of Jean Charles de Menezes in the Stockwell station of the London Underground is starkly illustrative of what is at stake. Indeed, this incident provides a tragic demonstration of the potential consequences of adopting preemptive security strategies and the concomitant compression of the decisional moment that this implies. To briefly recount the incident, in the frenzied aftermath of the 7/7 London bombings, Menezes, a Brazilian electrician living in the United

Kingdom on an expired student visa, was erroneously believed to be a prospective suicide bomber and was preemptively shot multiple times and killed in accordance with the London Metropolitan Police's "Operation Kratos" anti-terror policy (Vaughan-Williams 2007). This policy authorized armed officers to fire at the heads of suspected suicide bombers without warning if the detonation of a bomb was perceived to be imminent by the senior agent on the scene. The rationale behind the policy was to provide officers with the prerogative to "neutralize" a potential threat without accidentally detonating an explosive device located on the suspect's person (Taylor 2006). On the day in question, a suicide attack was perceived to be imminent, and so interventionary action aimed at preempting the threat was immediately authorized under these guidelines. This resulted in the death of an innocent man at the hands of the pettily sovereign agents of the state precisely because, in their belief that an attack was imminent, the shooters did not take the time to assess the circumstances and instead undertook a preemptive strike based on an imagined future in which Menezes *had* detonated a bomb on the Underground.

This case is particularly instructive because on the one hand, the logic of preemption was clearly operative—as the Kratos policy was explicitly formulated to permit officers to intervene *before* a catastrophic detonation occurs—while on the other, the temporal compression inherent therein is clearly demonstrated by the fact that Menezes was denied adequate time to demonstrate his innocence by the officers' feeling impelled to make the "lightning decision" to anticipatorily shoot to kill. In other words, the overwhelming imperative to act *now* in order to preempt a catastrophic potentiality ultimately precluded the shooters from evaluating the situation in an adequate manner, since doing so would have revealed the erroneous identification of Menezes as a suicide bomber and he would not have been fatally shot. That the incident reflected a conspicuous "lack of time" given to Menezes is emphasized by the words of his cousin, who, when interviewed after the incident, pointedly claimed that the police "judged my cousin and sentenced him, all in the space of a moment" (quoted in Vaughan-Williams 2007, 187). As Vaughan-Williams puts it, "time was quite literally 'taken away' from [Menezes]" (187).

It is also worth noting, however, that while the logic of preemption was a necessary condition for the Menezes shooting—in that the incident would not have occurred as it did had the Metropolitan police not been operating under a preemptive strategy—it alone was not sufficient. Indeed, recalling the earlier distinction between the conceptual "logic" and practical "politics" of preemption, it should be recognized that the ultimate decision to shoot to kill in this case also resulted from Menezes's conformation to an imagined representation of threat in terms of both behavior (e.g., fleeing from police) and appearance (e.g., dark hair and complexion). This once more highlights the importance of such factors as racialized othering practices that are embedded in the broader politics of threat construction to the practical operation of preemptive security strategies "on the ground." But again, my chief concern is to theorize the politics of preemption in the abstract; thus the key point is that the temporal compression inherent to a preemptive politics is clearly illustrated by the Menezes case. In particular, the incident highlights how such an approach to (in)security governance makes it much more difficult to "take time" in the context of what are often life-and-death decisions. Thus, while this temporally compressive tendency is manifestly antidemocratic in theory, the Menezes killing shows how it can also have tragically violent consequences in practice.

AFFECTIVE POLITICS

The Menezes shooting also suggests the importance of further unpacking the role of affect in the operationalization of a preemptive politics. In this respect, recall the argument of the preceding section, based on Fisher's theoretical framework, that a radically uncertain future only becomes actionable in the present if the passions can be activated with respect to the wholly unknown/unknowable—a circumstance that requires a degree of temporal manipulation whereby our relation to the future is modified by a reformulation of the notion of imminence to effectively include any potentiality. While making clear that affective responses are key to triggering anticipatory action, these points only imply that this is because affective responses are a necessary component of *all* human decision-making—albeit particularly so with respect to explicitly future-oriented decisions concerned with planning for the future. Little is said about either the particular character of the affective responses through which a politics of preemption operates or the precise ways in which anticipatory decisions *necessarily* rely upon the play of affect. A more detailed exploration of these questions is required to complete the task of understanding how preemption operates as a political rationality.

The key point in this respect is anticipatory decisions are made in a compressed timescape, and thus they are often devoid of democratic deliberation; as a direct consequence, they are also often made primarily on the basis of fundamentally non-rational "gut feelings" (de Goede 2012; Elmer and Opel 2008, 14). The Menezes shooting is again quite instructive in this respect in that, confronted with a potentially imminent catastrophe and thus deprived of the time to adequately collect and consider the evidence, the most plausible explanation for the officers' decision to shoot to kill is that such a decision was based primarily on the *feeling* that an attack was in the offing—a feeling undoubtedly attributable to the still-fresh traumatic memory of the recently executed 7/7 bombings.[6] It is crucial to recognize that this is a direct result of the temporally compressive pressures inherent to the logic of preemption. Indeed, as Elmer and Opel emphasize, "gut instinct" emerges as the primary basis for action "in the absence of either time to contemplate decisions or adequate intelligence and research on which decisions are based"—a condition that describes both the context of a preemptive security politics in general and the particular circumstances that prevailed on the day of the Menezes shooting (2008, 14).

This importance of "gut feeling" or "instinct" perhaps most clearly illustrates how a politics of preemption operates through the affective realm, as it shows how affect is crucial to the decisional logic through which such a politics is practiced. In other words, affective resonances are not merely crucial to making the future actionable in the present—which was the point made in the above discussion of Fisher's framework—they also provide the underlying informational basis for such interventions in the context of a preemptive political rationality. Massumi's concept of the "affective fact" usefully captures this idea (see Massumi 2005, 7–8; 2010, 54–55). For Massumi, the idea of the affective fact refers to a condition in which an "affective mechanism . . . exhibit[s] the certainty to which empirical facts aspire"—that is, an empirically ungroundable set of premises to which the powerful resonance of affective response grants the semblance of veracity (Massumi 2005a, 7; 2010, 55). This idea of the affective fact describes precisely the sort of imagined potentialities upon which anticipatory action is premised, since the latter cannot be grounded in empirical data—as they exist only in the ultimately unknowable future—and yet are conferred with an adequate degree of veracity to serve as the basis for concrete interventions in the present (Massumi 2010, 68). A more basic version of this notion entered the popular cultural lexicon in the mid-2000s—a point, not coincidentally, when preemption was becoming consolidated as a powerful political rationality in Western polities—through political satirist Stephen Colbert's coining of the term "truthiness" during the debut broadcast of his highly popular nightly program *The Colbert Report*. Using it in part to describe the George W. Bush administration's approach to policymaking, Colbert defined "truthiness" as "truth that comes from the gut, not books," and the term was ultimately selected as Merriam-Webster's 2006 Word of the Year, suggesting its utility for describing a significant element in contemporary public discourse (Merriam-Webster 2006).

Returning to the particular context of preemptive politics, affective facts can be understood as both the "knowledge" produced through the affective responses triggered by temporal manipulations which cast imagined catastrophes as imminent and the corresponding gut feelings regarding what intervention must therefore be made. In other words, affective facts are the collection of affects that comprise the lens through which a perceived threat is viewed and ultimately provide the informational/epistemic basis for a decision regarding the proper course of anticipatory action. Indeed, since preemptive governance strategies are inevitably burdened by "the decline of the empirical fact"—as a result of their attempt to act upon the unknowable future in a context of radical uncertainty—the affective fact serves as the primary basis upon which anticipatory action is mobilized (Massumi 2005a, 7). The affective fact is crucial to the logic of preemption, since it fills the epistemic void created by both the irrelevance of empirical facticity in the face of a radically uncertain future *and* the lack of time for deliberation about appropriate courses of action—both of which, as we have seen, are part and parcel of a preemptive politics. The primary epistemic basis for anticipatory decisions is therefore derived from what William Connolly terms the "visceral register"—specifically in the form of affective facts consisting of speculative imaginings of potential futures and gut feelings about proper courses of action (Connolly 2002, 130).[7]

The aim of this section has been to reflect upon a pair of additional considerations relating to the broader argument that a preemptive politics operates through a particular process of temporal manipulation that expands the "imminent future" indefinitely. In this regard, not only does this produce a radically altered subjective relation to time and the future, but it also radically alters the context of political praxis—specifically by compressing the temporality of political decision-making to the extent that the compatibility of preemptive governance with the procedural and juridical norms of liberal democracy can be seriously called into question. Moreover, this compressed political timescape combines with preemption's underlying imperative to govern an ultimately unknowable future in a context of radical uncertainty to force anticipatory decisions to rely upon the "gut feelings" of deciding authorities—on "affective facts" rather than empirical facts—as the primary basis for action. These additional considerations are vital to understanding how the logic of preemption functions as a political rationality.

CONCLUSION

The proliferation of preemptive rationalities in the security context has led to a significant rearticulation of what the very idea of "security" is understood to mean. In particular, vibrant post-Cold War scholarly debates about proper referents and threats have effectively been reframed in a decidedly temporal way, as a radically uncertain and potential catastrophic future has been inscribed as that which must be secured against. The pursuit of this understanding of security requires a political rationality which allows human agents to act upon the future through anticipatory interventions in the present, and this chapter's discussion has attempted to develop a detailed conceptual account of the practical logics underpinning such anticipatory rationalities. In particular, the discussion cataloged how preemptive security governance relies upon a form of temporal manipulation through which our relation to the future is radically modified. Specifically, all future potentialities must be considered potentially imminent, since only then can a mere abstract potentiality trigger the sorts of affective responses necessary to make it politically actionable in the present. Building on these points, it was also argued that there is more to understanding how preemption operates than this conspicuous process of temporal manipulation. Two points are of particular note in this respect. The first is that the perpetual imminence of catastrophe that this temporal manipulation presupposes translates into a highly compressed timescape of political praxis. This dynamic of temporal compression militates against liberal democratic political and judicial norms, since the time for deliberation about available evidence and proper courses of action is overridden by an apparent need to act *now*. Relatedly, the second point is that preemptive modes of governance depend significantly upon the role of affect, in that the "gut feelings" of decision-makers are crucial to its practical operation. This is because the underlying imperative to act upon an inherently unknowable future, combined with a compressed political timescape, precludes the taking of action on the basis of a carefully considered weighing of empirically verifiable evidence. Rather, what Massumi calls "affective facts"—constructed on the basis of affective responses to mere potentialities—form the primary basis for preemptive decisions. In addition to explaining how the logic of preemption functions as a political rationality, these points provide a preliminary indication of the degree to which the adoption and implementation of anticipatory governance rationalities entails radical modifications to the way political power is organized and exercised, both in the security realm and beyond. It is to this theme that the next chapter turns.

NOTES

1. This particular point about the prioritization of the future, and its implications for the exercise of political power, are discussed at length at the beginning of the next chapter.

2. This was a turn of phrase deployed repeatedly by Bush administration officials when challenged to provide clear proof (the proverbial "smoking gun") of Saddam's WMD program, and pithily captures the essence of the preemptive ethos that underpinned its post-9/11 national security decision-making.

3. These points can also be related back to the themes of chapter 1, in that early modern political theorists argue that fear—specifically fear of violent death—is constitutive of sovereignty itself. Hobbes, for instance, famously claimed that a sovereign commonwealth comes into being "when men singly, or many together by plurality of voices, for fear of death or bonds, do authorize all the actions of that man or assembly that hath their lives and liberty in his power" (Hobbes 2011 [1651], 182). This suggests that the role of the passions in shaping the conduct of political activity in the sense described by Fisher has long been recognized.

4. This terminology suggests an allusion to Nietzsche's vivid account of the emergence of the state itself, which he sees as resulting from the forceful action of those "who [are] by nature 'master,'" in lieu of "that sentimentalism which would have it begin with a contract." Of the former, Nietzsche writes, "They appear as lightning appears, too terrible, too sudden, too convincing, too 'different' even to be hated" (Nietzsche 1989, 86). Massumi's allusion is certainly deliberate since, as will become especially clear in the next two chapters, Nietzsche's account offers an apt description of the exercise of sovereign power in the context of a preemptive security regime. I thank Peter Nyers for bringing this connection to my attention.

5. The antagonistic relationship between a politics of preemption and the norms of a liberal juridical order will be explored at much greater length in the next chapter.

6. This point highlights the importance of questions of memory and trauma to the affective temporalities of preemptive security practices. While an exploration of such questions is beyond the scope of this chapter, it is important to recognize that the affective resonances operating within a politics of preemption will inevitably be conditioned by memoraic traces of traumatic experience. On this issue more broadly, see the work of Maja Zehfuss (2003) and Jenny Edkins (2002).

7. The next chapter returns to this issue in its consideration of how the construction of the epistemic/evidentiary basis for preemptive action relies upon the exercise of the imagination. While the arguments are not couched specifically in terms of the role of affect and/or affective facts, it should be kept in mind that the process of imaginary exercise that I do discuss refers to a central aspect of the production of the affective facts discussed here as the legitimative foundation for preemptive action (Massumi 2010, 54).

Chapter Five

Preemptive Security and the Politics of Exceptionalism

THE ATTORNEY GENERAL SPEAKS

On March 5, 2012, US Attorney General (AG) Eric Holder delivered a speech at the Northwestern University Law School in Evanston, Illinois, outlining the Obama administration's legal justification for the element of its counter-terrorism strategy, under which suspected terrorists are targeted and killed by missile strikes from unmanned aerial vehicles, more commonly known as "drones" (United States Department of Justice 2012).[1] Begun as a more covert element of the Bush administration's anticipatory approach to the post-9/11 governance of transnational terrorism (Kessler and Werner 2008), this so-called Killing Program, carried out primarily by the Central Intelligence Agency (CIA), has been expanded under Obama, to the extent that it now constitutes perhaps the central component of American anti-terror efforts (Leander 2011).[2] Holder's speech was specifically occasioned as a response to growing concerns from civil libertarians about the program, which reached a peak after the September 30, 2011, killing of Anwar al-Awlaki—a radical Islamic cleric with ties to al-Qaeda who was also an American citizen. This latter point in particular worried critics, who argued that the killing of a US citizen by the American government in this manner constituted a dangerously flagrant violation of the Due Process Clause of the Fifth Amendment to the US Constitution (see Greenwald 2012). Holder's speech aimed to counter such charges by elucidating a legally justificatory rationale for using targeted killing as a key tool in the global fight against terrorism—even if those targeted happen to be citizens of the United States. While the jurisprudential cogence of the administration's arguments can certainly be called into question, I leave such questions to legal scholars (Blank 2015).

For the purposes of this chapter, however, the speech is notable for the way its defense of the administration's policy explicitly invokes a pair of tropes that have become central to (inter)national security discourse in the decade since the advent of the so-called War on Terror. These tropes are the imperative of preemption—which has been discussed at length thus far—and the idea of "exceptionalism"—which has been hinted at in passing and will be unpacked at length below.

With respect to the first of these tropes, Holder emphatically reaffirmed the Bush-era embrace of the logic of preemption as the basis for the US approach to governing terrorism. Arguing that "al-Qaeda has demonstrated the ability to strike with little or no notice—and to cause devastating casualties," he claimed that the president must not be required "to delay action until some theoretical end-stage of planning—when the precise time, place, and manner of an attack become clear, [because] such a requirement would create an unacceptably high risk that our efforts would fail, and that Americans would be killed."[3] This line of thinking is echoed in a Justice Department White Paper on targeted killing that was leaked to the press some months after the speech, which argues that the identification of an "imminent threat" against which preemptive action can be legitimately taken "does not require the United States to have clear evidence that a specific attack on US persons or interests will take place in the immediate future" (United States Department of Justice 2013, 7). What Holder and the Obama Justice Department are arguing, is that in the face the potentially catastrophic effects of the next attack, efforts to govern terrorism must adopt an overtly anticipatory approach informed by a precautionary logic, whereby interventions such as targeted killing are undertaken despite a low level of certainty regarding the nature or imminence of the threat posed by those being targeted. Near the conclusion of the speech, Holder articulates this perspective more succinctly, asserting that, "[i]n this hour of danger, we simply cannot afford to wait until deadly plans are carried out—and we will not."

While clearly and explicitly invoking the post-9/11 conceptualization of preemption that was discussed at length in chapter 3, Holder's language here—particularly his allusion to the "hour of danger"—also hints at the second trope mentioned above: that of political exceptionalism. Indeed, throughout the speech, Holder implicitly deploys a logic of exceptionalism as the basis for his arguments about the legitimacy of targeted killing, particularly when those targeted are US citizens. Generally speaking, the idea of exceptionalism refers to a strengthening of executive authority—ostensibly in response to an acutely dangerous situation—that results in "serious distortions in the restraining effects that the rule of law and democratic representation have on the arbitrary exercise of power" (Huysmans 2004, 327). As articulated in the speech, Holder's defense of the targeted killing of citizens by the US government is rooted in precisely this idea, as he asserts that the capacity to properly confront the threat of terrorism "depend[s] on expertise and immediate access to information that only the Executive Branch may possess in real time." Accordingly, he further argues that "the Constitution's guarantee of due process . . . does not require judicial approval before the president may use force abroad against a senior operational leader of a foreign terrorist organization with which the United States is at war—even if that individual happens to be a US citizen." In this bold statement, Holder is explicitly claiming that the extraordinary exigencies of the War on Terror—and the attendant imperative to preempt potential terrorist attacks—demand the ultimate primacy of executive authority over the judicial checks on that authority inscribed in the Bill of Rights. Indeed, with such an eschewal of the juridical circumscriptions conventionally associated with the Due Process Clause, the only remaining limit to executive power in this context is the requirement that, as Holder put it, "The targeted individual poses an imminent threat of violent attack against the United States."

The inconsequentiality of such a limit should be readily apparent in light of the previous chapter's discussion, since the understanding of preemption that informs the post-9/11 governance of (in)security—to which Holder's remarks clearly conform—involves the framing of *any* potential threat as "imminent" for the purposes of triggering anticipatory action. However, in the particular case of the targeted killing program, the shortcomings of this limitation on executive power are further exacerbated, since the final decision as to whether the proper threshold of imminence has been met *also* rests with the executive alone. As former US Defense Secretary Leon Panetta described the process in a January 2012 interview with *60 Minutes*, "The president of the United States . . . reviews these cases, reviews the legal justification, and in the end says, go or no go" (CBS 2012). In other words, the life-and-death decisions associated with the drone program are deemed to be the sole purview of the president himself; thus, they are not subject to independent assessment by a federal court or any other element of the judicial branch. Such executive skirting of institutional legal constraints and norms, all in the name of taming the radical contingencies attributed to the current security moment, represents an archetypical manifestation of political exceptionalism—a point perhaps best captured in what became the takeaway sound-bite of Holder's speech, in which he claimed that "'due process' and 'judicial process' are not one and the same, particularly when it comes to national security."

PREEMPTION AND EXCEPTIONALISM

This chapter began with an analysis of Attorney General Holder's remarks because they provide a useful empirical frame for the more abstract conceptual questions that will be taken up below. In particular, the speech and its context illustrate two points that serve as the basis from which this chapter's discussion will proceed. The first point is that the ongoing juridico-political debate around the drone program in general, and its targeting of American citizens in particular, demonstrates that despite nearly a decade having passed since the Bush administration altered the global security imagination by enshrining preemptive self-defense in its National Security Strategy—and despite the fact that the associated "Bush Doctrine" of preemptive inter-state war has been largely discredited by the imbroglio that resulted from its pursuit in Iraq—the logic of preemption, as such, remains a powerful political rationality with respect to the governance of transnational terrorism. Indeed, Holder's explicit invocation of a preemptive imperative suggests that the reorientation of the global security imagination along anticipatory lines continues apace. Second—and more important for the purposes of this chapter—Holder's invocation of political exceptionalism as a legitimating rejoinder to the apparent extrajudiciality of the al-Awlaki case in particular suggests that the operational practicalities of preemptive security are intimately bound up with an exceptionalist politics characterized by the suspension of the prevailing juridical order, the emergence of a "decisionist" paradigm of sovereignty, and the subsequent creation of a condition in which this sovereign power "confronts . . . life without any mediation" (Agamben 2000, 41). Granting the executive the capacity to circumvent judicial restraints when deciding upon the killing of a citizen is an archetypical manifestation of precisely such an exceptionalist politics. This suggests an intimate connection between the logic of preemption and the politics of exceptionalism—a connection whose interrogation will be a primary concern of this chapter, and which, in turn, illustrates the broader point that a politics of anticipatory governance involves a radical reconfiguration of how political power is organized and exercised that challenges the constitutive norms of liberal democratic polities.

Highlighting a connection between preemption and exceptionalism is no new revelation, of course, in that much recent critical literature has drawn attention to such a link to varying degrees and in varying ways (see Amoore 2014a; Aradau and van Munster 2009, 2011; de Goede 2011, 2012; Ericson 2008). However, this empirically rich and theoretically sophisticated body of work has largely limited its analysis of the preemption/exception intersect to observing and/or demonstrating that specific anticipatory security practices across the global War on Terror embody certain characteristics of the paradigm of exceptionalism as theorized by Giorgio Agamben via Carl Schmitt. While these analyses are certainly important, such a primarily empirical and correlational understanding of the connection between preemptive security and political exceptionalism does not fully capture its originary depth. In other words, the existing literature demonstrates conclusively *that* such a connection exists without significantly fleshing out the conceptual reasons *why* this is so. This is the question that this chapter seeks to explore. In this respect, it is argued that this correlation between preemptive governance and exceptional politics is no coincidence, since political exceptionalism can be understood as logically presupposed by the sorts of preemptive political rationalities that characterize the global War on Terror. Moreover, in accordance with this book's broader commitment to employing a temporal analytical lens, I will consider these questions by engaging with an aspect of the politics of preemption that has, perhaps somewhat surprisingly, been granted relatively little attention in the scholarly literature: the question of its constitutive temporality. This chapter's attempt to think through the connection between preemption and exceptionalism will be rooted in an explicitly temporally inflected reading of the former, as I will contend that the exceptional practices associated with preemptive security are directly traceable to the logic of preemption itself, and this link is ultimately a function of the unique political temporality presupposed thereby. It is in this capacity that the practical correlation between preemptive security and exceptionalism is something of a logical necessity stemming from the conceptual constitution of the logic of preemption as a political rationality.

This may initially seem like an ambitious attempt to develop a theoretically complex line of reasoning. Yet, in light of AG Holder's remarks and the still-ongoing debate over American drone warfare and other anticipatory security measures and strategies, such conceptual questions are far from irrelevant. Indeed, given the enduring global importance of preemptive security rationalities, the continued proliferation of political exceptionalism, and the relatively limited theorization of preemption's political temporality in the existing literature, I believe that an attempt to untangle the conceptual relationship between preemption and exceptionalism through a temporally oriented analytical lens can offer a crucial contribution to efforts aimed at developing a thorough conceptualization of the contemporary landscape of global (in)security governance in particular, and the broader rise of anticipatory forms of societal governance more generally.

With these points in mind, then, the chapter will proceed in three subsequent sections. The first will examine and explain what might be called the political temporality of preemption. Here it will be argued that when adopted as a rationality for societal governance—particularly in the security context—the logic of preemption implies a political paradigm in which the governance of the future is normatively prioritized, while the present is concomitantly instrumentalized as the location of the disruptive anticipatory interventions required to render this priority actionable. I will then consider how such a reformulation of the present/future relation enacts a logic of political action premised upon speculation (de Goede 2012), conjecture (Aradau and van Munster 2011), and the exercise of the imagination (Salter 2008a), which in turn skews the practice of politics in a decidedly exceptional direction. This claim provides the foundation for the remainder of the chapter, and will be subsequently unpacked at length. However, prior to proceeding in this direction, the second section takes a detour into an account of precisely what is meant by "exceptionalism" in the present context. While acknowledging the myriad approaches to this notion in recent scholarship, I contend that the idea of exceptionalism is fundamentally characterized by two core components, or "pillars": the suspension of the juridical order, and the enaction of a "decisionist" form of sovereign power. The remainder of the chapter builds on this point to argue that the operationalization of preemptive security rationalities in particular requires a paradigm of political authority and sovereignty that conspicuously embodies these two elements. This claim is elaborated at length in the third section, where I engage in a more detailed exploration of each pillar of exceptionalism, emphasizing how the enaction of each is presupposed by a politics of preemption. While the essence of this discussion is theoretical, the core conceptual claims are fleshed out with extensive reference to illustrative examples taken from the ongoing prosecution of the War on Terror. The chapter concludes by taking stock of the argument and preliminarily considering the implications that follow from the originary conceptual relationship between preemption and exceptionalism that has been elucidated. This lays the foundation for the next chapter, which will consider these implications in more detail, particularly as they relate to experiences of political subjectivity.

FUTURITY AND IMAGINATION

Understanding the conceptual relationship between preemptive governance and political exceptionalism must begin by considering what can be understood as the political temporality of preemption—meaning how time is integrated into the logic of preemption in its capacity as a political rationality. Continuing with the theme of this chapter (and part II of the book more broadly), this idea can be usefully unpacked through a closer examination of preemption as applied to the post-9/11 governance of (in)security. In this respect, it is helpful to begin by recalling why preemption has proliferated as a security rationality in the post-9/11 era, since its political temporality is directly related to the particular type of security problem(s) to which it has emerged as a response. Returning to the discussion in chapter 3, the key here is that the rise of preemptive security strategies has corresponded with the emergence of transnational terrorism as the dominant global security concern, since the threat of terrorism has been articulated in overtly temporal terms. Indeed, as we have seen, the dominant framing of the 9/11 attacks as a radically irruptive "event"[4] led to the specific security problem of terrorism becoming identified with the broader existential problem of taming temporal contingency. This framed terrorism as something whose governance required asserting a degree of control over the unfolding of a radically uncertain future, and the post-9/11 articulation of preemption is premised upon granting precisely such capabilities to those who deploy it as a governance strategy. Thus, with the adoption of preemptive rationalities, the emphasis of (in)security governance shifts from the realm of the spatial—where defense of the sovereign frontier from armed incursion is prioritized—to the temporal—where precluding catastrophic irruptions potentially lurking in the uncertain future is the ultimate aim.

This reorientation of (in)security governance implies a very particular political temporality, which can be distilled down to one core idea: *the prioritization of the future over the present.* To understand this point—which is key to the conceptual link between preemption and exceptionalism—we must first recognize that the very notion of preemption as a political rationality is rooted in a particular normative claim—namely, that in the face of a radically uncertain, potentially catastrophic future, a perspective that "require[s] the catastrophic prospects of the future be avoided at all costs" represents the only ethically legitimate orientation for societal governance (Aradau and van Munster 2009, 695; 2011, 42). Anna Leander articulates this idea well: "The future cannot be known . . . but if a risk is imminent it would be irresponsible not to take the prudent, preventive, and/or precautionary measures necessary to protect oneself (and others) from it" (2011, 2256). We have previously seen how this perspective translates into an overarching imperative that some sort of intervention must be made to actively diminish, if not eliminate, the potential for an imagined catastrophic future to emerge, regardless of the degree of (un)certainty with which the relevant decision-makers are confronted. Adopting preemption as the basis for (in)security governance implies the acceptance of this imperative and a commitment to act accordingly. The corollary of this commitment, however, is that a particular relationship between present and future is enacted, whereby the former is effectively subordinated to—or even "taken hostage" in the name of—the latter in the context of political decision-making and action (Stockdale 2013, 153).

This can be illustrated by the very basic point that any practical intervention aimed at acting upon the future must necessarily take place in the present. The simple logic of temporal causation affirms that any act seeking to create a future effect must temporally precede the desired moment of influence, and, as beings in time, human agents are only capable of taking action in the lived present where they continually exist. Thus, since anticipatory interventions aim to influence the future on the one hand, and constitute acts of human agency on the other, they must necessarily take place in the present. It therefore follows that in the security context, because a preemptive approach requires that some sort of action be taken to govern the future, and because such action must necessarily take place in the present, the ontology of security is articulated in such a way that the *future itself* effectively constitutes the referent to be secured, while the present is *instrumentalized* as the necessary location of anticipatory interventions required to pursue this end. In other words, the logic of preemption prioritizes the governance of the future, which in turn implies the legitimation of potentially disruptive interventions in the lived present, effectively placing the latter in subordinate service to the former. It is in this respect that the logic of preemption can be understood to take the present hostage in the name of the future.

A non-security example can help further illuminate these points, as preemption's subordination of present to future is analogous to the logic underpinning the doctrine of fiscal austerity that has driven economic policymakers seeking to confront the sovereign debt problems that emerged in the aftermath of the 2008 global financial crisis. In this respect, austerity proponents have asserted that national economies must endure painful contractions of government spending on social welfare programs in the present as a necessary means for adequately securing the future from financial instability caused by bond market hostility to an excess of sovereign debt. In a similar manner to the logic of preemptive security, therefore, the present is instrumentalized as the location of highly disruptive and potentially damaging action that is nevertheless deemed necessary for the realization of a future in which imagined crises are precluded. It is in this sense that we can speak of preemption (and austerity) as involving the prioritization of the future over the present, since the present must absorb the interventionary blows that comprise any attempt to secure the future from its own potentially catastrophic contingencies.

The importance of this normative prioritization of the future—and the attendant taking of the present hostage—is that it has the effect of altering the epistemic foundations of political action in a highly significant way. The key point, here, is that such future-prioritizing political rationalities require that decisions be "taken beyond the realm of certainty" (Aradau and van Munster 2008, 32), thus implying a decisional logic that relies primarily upon "knowledge" derived from the exercise of the imagination.[5] Indeed, as Claudia Aradau and Rens van Munster assert, "imagination acquires epistemic primacy in relation to the unknown" (2011, 85), and because preemption seeks to act upon an uncertain future that is by definition unknowable, the exercise of the imagination is vital to rendering the proverbial next attack—or any future contingency to be preempted—politically intelligible and thus anticipatorily actionable in the present. In the particular context of preemptive security, therefore, "imagination is constitutive of security knowledge" (84), and "the use of imagination as a supplement to calculus as a modality of anticipation of the future has been encouraged" (Bonelli and Ragazzi 2014, 478). Any decision regarding whether and how anticipatory action is taken will rely upon the imagination to a significant degree, suggesting that a politics of preemption shifts the primary epistemic basis for action from the realm of empirically verifiable fact to the realm of speculation, conjecture, and suspicion (Anderson 2010b, 228; Aradau 2015, 157–58; de Goede and Randalls 2009, 868; Salter 2008a, 243).[6]

Crucially for the purposes of this chapter, this shift also implies a significant modification to the way political power is exercised, since it suggests that anticipatory action is based to a significant degree on an evidentiary basis that must be *actively constructed* by relevant authorities. In other words, under a politics of preemption, those tasked with making anticipatory decisions are *also* tasked with creating the very epistemic foundation for these decisions through the exercise of the imagination. The preceding chapter offered an abstract theorization of how this process operates, emphasizing the importance of "gut feelings" and "affective facts" in this context (Massumi 2010, 2005). I will not revisit this discussion here, as the key point for present purposes is simply that anticipatory interventions carried out under a politics of preemption effectively constitute "ungrounded, arbitrary attempt[s] to subdue the contingency of the future" (Aradau and van Munster 2008, 32), which has the crucial effect of enhancing the scope of decisional authority granted to those tasked with deciding how and when to act by requiring them to actively create the epistemic basis for action.

These conceptual points can be clarified through an illustration from the prosecution of the global War on Terror. On March 29, 2011, Mohamed Hersi, a twenty-five-year-old Canadian of Somali descent, was arrested by the

Royal Canadian Mounted Police (RCMP) at Toronto's Pearson International Airport as he tried to board a flight to Cairo via London. After being charged and tried, Hersi ultimately became the first Canadian to be convicted of "attempting to participate in a terrorist activity"—an inchoate offense added to the Criminal Code of Canada in the aftermath of the 9/11 attacks—for which he was sentenced to the maximum of ten years in prison (CBC News 2014). Like most Western states in the post-9/11 era, Canada adopted a preemptive approach to governing terrorism premised upon the goal of "protecting Canadians from terrorist acts *before they occur*" (Government of Canada 2011, 32, emphasis added). The country's anti-terrorism guidelines have been developed with the aim of allowing security agencies to act in such a way that threats can be "dealt with on more of an *a priori* basis rather than more of a post facto basis" (Svendsen 2010, 320). This explicitly anticipatory outlook has been described as "risk preemption"—an interesting turn of phrase, which suggests that the ultimate aim is not merely to preempt already existing threats, but to preclude even the possibility that such threats may develop at some point in the future—that is, to preempt risk itself (307). The RCMP was operating under this ideational regime when Hersi was detained.

Despite his insistence that he merely planned to study Arabic in Egypt for four months, the RCMP alleged upon his arrest that Hersi intended to proceed to Somalia to train as a militant with the Al-Shabaab network—a group classified as a terrorist organization under Canadian law, whose most high profile act to date has been the September 2013 attack on the Westgate shopping mall in Nairobi. At the time of Hersi's arrest, however, the RCMP offered a strikingly candid admission similar to that of the FBI in the paid informant case discussed in chapter 3, in which it acknowledged that there was very little concrete evidence that Hersi would, in fact, proceed to do what they alleged he was planning to (Freeze and MacArthur 2011). Indeed, it was conceded that the RCMP's investigation "did not indicate that the suspect was a direct threat to his country or Canadians" at the time of his detention (Teotonio 2011). The threat posed by Hersi that was preempted by his arrest did not tangibly exist at the moment he was detained, but was instead located in an imagined potential future that may or may not have ever come to pass, as Hersi was readily acknowledged to have been several degrees removed from actually posing what would conventionally be considered an imminent threat. This aspect of the case was used unsuccessfully in Hersi's defense at trial, after which his lawyer promised to appeal and claimed with deliberately chosen words that the conviction relied on conjectural evidence that was "so far removed from reality as to make it a thought crime" (Humphreys 2014).

Hersi was thus subjected to an intervention by agents of the Canadian state based on an *imagined future threat* that he *may have one day* posed had he ultimately succeeded in acting upon what he was *suspected* of *planning* to do. The basis for his arrest thus amounted to a series of speculative assertions, as the relevant authorities relied upon a primarily conjectural construction of the future when deciding to preemptively detain Hersi in the present. It should also be recognized that such speculative imaginings as are necessary for the practice of preemptive security do not take place in a vacuum, since the process of conjectural knowledge creation, upon which anticipatory interventions rely, inevitably reflect broader discourses of threat relating to particular types of individuals that have been cultivated and circulated in the prevailing security climate. The Hersi case exemplifies this point as well, in that his arrest came amidst growing concern in Canadian national security circles about the "radicalization" of young Muslim men in the Somali-Canadian diaspora (Canadian Security Intelligence Service 2011). While it is crucial to recognize this element of preemptive security, my concern here is less with the substantively racialized orientation of the contemporary Western security imagination, and more with how the logic of preemption requires the mobilization of that imagination for the purposes of anticipatory action. The Hersi case is thus quite instructive, since the RCMP were operating under an explicitly preemptive rationality prescribed by the Canadian government; as such, a reliance on speculation did not prevent them from intervening against Hersi. The case is also instructive because the RCMP were acting as precisely the sorts of "petty sovereigns" that were described in chapter 3 as often being on the front lines of a preemptive security regime. Indeed, the RCMP made the ultimate decision to preemptively arrest Hersi, but they were also operating under the broader governmental umbrella of the Canadian state's preemptive national security policy as set by the prime minister and cabinet (Government of Canada 2011). In short, Hersi's arrest and conviction provide a useful illustration of how the praxis of preemptive security privileges the imagination as the epistemic foundation for action, and thus expands the decisional authority of pettily sovereign security agents by requiring them to conjecturally construct that foundation.

To conclude this section, it should be acknowledged that the importance of the imagination to the decisional logic of preemptive security has been highlighted in various ways by other critical scholars of security. Mark Salter, for instance, has asserted that "[t]he logic of preemption prioritizes the power of imagination over the power of fact" (2008a, 243), while Aradau et al. affirm that "the sovereign order is no longer simply that of decision, but also that of imagination" (2008, 152). Yet the existing literature has not considered in any great depth how this reliance upon speculative knowledge modifies the way political power is exercised in such a way that reveals a crucial *conceptual* relationship between the logic of preemption and the politics of exceptionalism. This is the focus of the remainder of this chapter, wherein I argue that the aspects of preemptive security discussed in this section—namely, a political temporality that prioritizes the future over the present and thereby increases the scope of decisional authority by placing the imagination at the core of the sovereign decision—demonstrate that the correlation between the rise of preemptive strategies and the prevalence of "exceptionalist" practices in the global governance of (in)security is no coincidence, since the latter follows logically from the former. Before turning to this line of the argument, however, a brief discussion of precisely what is meant by the term "exceptionalism" is required.

THEORIZING POLITICAL EXCEPTIONALISM

The idea of political exceptionalism—and the related notion of the (state of) exception—has become perhaps the primary theoretical frame for the critical study of security in the post-9/11 period. Indeed, since the advent of the War on Terror, a voluminous literature has emerged in which the concept of the exception is both unpacked conceptually (Huysmans 2004, 2006, 2008; Lazar 2006, 2009; Prozorov 2005; Strong 2005; Walker 2004, 2006) and deployed as a lens through which to theorize a wide range of specific practices undertaken in the name of governing transnational terrorism and associated issues of (in)security (Amoore 2008; Aradau and van Munster 2009; Ericson 2008; Nyers 2006; Salter 2008b; van Munster 2004). Often invoking the thought of the two figures most closely associated with the idea—Giorgio Agamben (1998, 2005) and Carl Schmitt (2005)—such work has emphasized the analytical utility of the concept of exceptionalism for developing critical readings of the rationalities, strategies, and tactics employed by sovereign authorities in the security context. While generally sympathetic to this line of inquiry, I want to engage with the idea of political exceptionalism in a somewhat different way. My aim is not to further demonstrate that "exceptionalist" modes of governance pervade the contemporary security landscape—a task that has been thoroughly undertaken by the texts cited above, among many others—but rather to emphasize that perhaps the dominant security rationality of the current moment, the logic of preemption, can only be operationalized through practices best described as exceptionalist. It is necessary at this juncture to briefly unpack precisely what is meant by the notion of political exceptionalism. While the following discussion can in no way do justice to the full range of highly nuanced and theoretically sophisticated articulations of exceptionalism developed in the existing literature, there are certain core features of the concept that can be elucidated and serve as a foundation for the subsequent arguments.

To begin most basically, the idea of exceptionalism should be understood as being located at the interstice of the political and the juridical. In particular, it is concerned with the limits of the rule of law regarding the exercise of sovereign political power, as an "exceptional" condition is said to emerge when the former is suspended and latter is thus no longer circumscribed by any prevailing normative strictures (Huysmans 2004, 327). In other words, a politics of exceptionalism exists when the legal order is annulled and an ultimately arbitrary mode of power, effectively unconstrained by law and most often vested in the person or office of the executive, emerges as the primary node of governmental authority. As Carl Schmitt puts it in his seminal text *Political Theology*, "[W]hat characterizes the exception is principally unlimited authority, which means the suspension of the entire existing order . . . the state remains, whereas the law recedes" (2005, 12). Historically, exceptionalism has been associated with a particular political circumstance in which the very existence of the polity is perceived to be under threat "from some grave internal or external danger" (Lazar 2009, 20; Nyers 2006, 27). The idea possesses a rich historical lineage in both political and jurisprudential theory, as the conceptual kernel of exceptionalism can be traced from modern liberal constitutional articulations of "emergency powers," through late medieval and early modern conceptions of the "state of siege," back to antiquity and the notion of the "iustitium" upon which the concept of the Roman dictatorship was premised (see Agamben 2005, 11–22, 41–51; Lazar 2009, 113–35). While the congruence of such provisions with contemporary norms of liberal democracy has been a vociferously debated question among political and legal theorists—particularly since the state of emergency/exception-type argument has been regularly deployed to legitimate numerous anti-terror policies of dubious constitutionality[7]—the general conception of exceptionalism to which such debates refer has remained relatively constant, signifying a condition "in which decision-making cannot be subsumed by existing [legal] norms . . . in which norms are held open to suspension or transformation, and where programs of norm-implementation and norm-compliance cease to govern action and decision-making" (Johns 2005, 619). In short, exceptionalism refers to a political condition in which the executive—and, importantly, the petty sovereigns that often act as its surrogates[8]—is vested with an authority unconstrained by the rule of law and reminiscent of that of the absolute sovereigns posited by a tradition of political theorists from Bodin to Schmitt. It is thus to such an understanding of the term that this chapter refers when speaking of political exceptionalism.

Much scholarly energy has been expounded theorizing exceptionalism. And while it is certainly important to understanding the concept, the resultantly complex debates concerning whether and how the idea of the exception is constitutive of all normative order—as per the so-called Schmittian idiom (Huysmans 2008, 180; Prozorov 2005; Schmitt 2005; Strong 2005)—and/or whether and how the state of exception has emerged as the dominant "paradigm of government" and thus "become the rule" in the current moment—as per the "Agambenian idiom" (Huysmans 2008, 180; Agamben 2005)—are of limited relevance here and will not be considered further. Instead, following from the general formulation of exceptionalism developed above, I will elucidate the concept itself in further detail so as to lay the foundation for the remainder of the discussion. In this respect, the notion of exceptionalism as theorized in the canon of political thought can be understood as characterized by two core components: the suspension of the juridical order, and the creation of a "decisionist" paradigm of sovereign authority. In addition to offering a parsimonious conceptual elaboration of the idea of exceptionalism on its own terms, such an understanding is useful for the purposes of this chapter, since it is with respect to these two "pillars" of political exceptionalism that the conceptual connection between the latter and the politics of preemption can be best elucidated. To begin moving toward this end, let us briefly consider each of these two pillars in turn.

The Two Pillars of Exceptionalism

The first pillar of political exceptionalism can be best described as the *suspension of the juridical order*, since the enactment of a state of exception entails the emancipation of sovereign authority from the normative circumscriptions imposed by "liberal constitutionalist" attempts to subjugate it to the rule of law (Lazar 2006, 260; Schmitt 2005, 11). The effect of this liberation of sovereign authority from its juridico-normative bonds is the *de facto* negation of the legal order with respect to the relationship between state and citizen, or sovereign and subject. Indeed, under a condition of political exceptionalism, not only is the primary node of political authority no longer limited by the rule of law, but the inhabitants of the polity in which this condition obtains are no longer insulated from arbitrary exercises of that power, since this protective task generally falls to the mediations provided by the law. As a consequence, the sovereign can visit what amounts to arbitrary violence upon targeted individuals or groups largely at will. This has the effect of placing political subjects in a "zone of indistinction," in which they are confronted by sovereign power as what Agamben has famously termed "bare life," which can be maimed or killed with impunity (Agamben 1998;

Edkins 2000). In the state of exception, in other words, the will of the sovereign supersedes the normative order of law and takes on the "force of law" itself (Derrida 1990). As Agamben puts it, political exceptionalism entails "the separation of 'force of law' from the law . . . on the one hand, the norm is in force but is not applied (it has no 'force') and, on the other, acts that do not have the value of law acquire its 'force'" (2005, 38). It is in this sense that the juridical order is suspended, as the declaration of the exception removes the sovereign's obligation to abide by the law in such a way that maintains the law's status as the overarching normative framework of societal governance. Accordingly, "the state of exception is . . . a space devoid of law, a zone of anomie in which all legal determinations . . . are deactivated" (Agamben 2005, 50).

In such a juridico-political environment, the exercise of sovereign authority takes on a decidedly purer form, since the absence of a legal order possessing the force of law invests the holders of that authority with an enhanced degree of decisional discretion. As Schmitt puts it, "[T]he decision frees itself from all normative ties and becomes in the true sense absolute" (2005, 12). This implies the second pillar of political exceptionalism: the enactment of *decisionism* as the prevailing form of sovereign authority. A decisionist paradigm accompanies the suspension of the juridical order that characterizes exceptionalism because it denotes a condition in which the sovereign does not simply apply the law as determined by a legislative body, but rather is "the *source* of law" (Lazar 2009, 36, emphasis original). In other words, when the juridical circumscriptions on the exercise of sovereign political power that are associated with the rule of law are removed through the declaration of the exception, every decision taken by the sovereign is necessarily purer in its absolute singularity, such that, "when looked at normatively, the decision emanates from nothingness" (Schmitt 2005, 32–33). In the vernacular of normative theory, the term "decisionism" refers to an approach to societal governance which espouses the virtues of a political system organized along such lines. The core of the decisionist argument is that such a paradigm of sovereign authority is necessary if any semblance of order is to be possible in the chaotic, contingent, and violent realm of human affairs (Lazar 2009, 36–42). While the details of this theory will be elaborated later in this chapter, the key point for present purposes is that a "decisionist" form of sovereign authority is a constitutive feature of political exceptionalism, since the elimination of the juridical constraints upon the exercise of sovereign power—and the concomitant conferral upon the sovereign decision the status of the force of law—brings into being a condition in which what amounts to a decisionist paradigm of sovereign authority obtains. It is in this sense that decisionism represents the second "pillar" of political exceptionalism.

PREEMPTIVE SECURITY AS EXCEPTIONAL(IST) POLITICS

While the degree to which the two pillars of exceptionalism have actually emerged in the contemporary global security climate is certainly debatable, there can be little doubt that myriad practices undertaken in the name of governing transnational terrorism evince such characteristics and can thus be considered "exceptional." Notably, some of the clearest examples in this regard are the sorts of explicitly preemptive practices with which this book is concerned, such as targeted killing (Kessler and Werner 2008), anticipatory asset freezing (de Goede 2011, 2012), and indefinite detention (Butler 2006; Ericson 2008). Accordingly, much of the diversely interdisciplinary scholarship on preemptive security in particular and anticipatory governance more broadly has frequently engaged with the concept of exceptionalism to a greater or lesser extent—so much so, in fact, that some of the most recent interventions have sought to move, in the words of Adey, Anderson, and Graham (2015), "beyond exceptionality" in their analyses. This suggests that the concept of exceptionalism has been largely exhausted as a productive theoretical framework for interrogating the politics of preemption and the governance of the future. There is perhaps some merit to such a claim, since the extant literature has quite comprehensively catalogued the extent to which the policies, practices, and strategies through which anticipatory governance rationalities are implemented often take an exceptionalist form. Yet such claims may still be somewhat premature, as there is an aspect of the well-documented correlation between preemption and exceptionalism that has yet to be fully considered—again, this is the extent to which the former logically presupposes the latter. What I try to show in the rest of this chapter is that the *practical* affinity between the logic of preemption and a politics of exceptionalism can, in fact, be explained at the purely *conceptual* level. In particular, it is argued that the prioritization of the future and attendant reliance upon the exercise of the imagination require that anticipatory approaches to (in)security governance be operationalized through a politics that effectively embodies the two "pillars" of exceptionalism outlined above. To unpack this claim, I proceed by considering these two pillars, in turn, with the discussions consisting of both a theoretical explanation of how each can be seen to follow from the logic of preemptive security and a discussion of illustrative examples taken from the War on Terror that further flesh out these conceptual claims.

The Suspension of the Juridical Order

We have seen that political exceptionalism is defined, in part, by the elimination of the normative constraints upon the exercise of sovereign power provided by the rule of law. Moreover, the extant literature on preemptive security has shown how such anticipatory practices often embody fundamentally extrajudicial elements (see de Goede 2011; Ericson 2008; Kessler and Werner 2008). As the following discussion will make clear, a reading of the logic of preemptive security that foregrounds its future-prioritizing political temporality reveals that these two points are intimately connected on a conceptual level. In particular, it shows that political action compatible with the imperative to govern an uncertain future is essentially *in*compatible with the normative requirements of a liberal legal order. Indeed, precisely because they seek to govern the future through sovereign interventions based primarily upon the exercise of the imagination, preemptive security rationalities suspend the juridical order at the very moment they are actualized in practice.

Conceptual Considerations

To elaborate on these points, recall that the logic of preemption is normatively premised upon taming the future's radical contingency, and that this imperative must be pursued through anticipatory interventions in the present framed in reference to imagined future events. As such, preemptive political rationalities embody a temporality in which the future is effectively prioritized over the present. However, because the future events it seeks to govern have not yet come to pass, and in fact may never do so, it also follows that the praxis of preemptive security must take the form of anticipatory action against individuals or groups based not upon a particular act they have committed, nor even upon what they have indicated through concrete actions or declarations that they intend to do, but rather on the basis of what it is imagined by the relevant authorities that they might (think about) do(ing) at some indefinite point in the future (Weber 2007, 115). Once again, the Mohamed Hersi case is illustrative here, as is the Obama administration's "watchlisting" program described in chapter 3, which federal court proceedings have shown operates through counterterrorism officials deciding that "people are not allowed to fly because of what the government believes they might do and not what they have already done" (Ackerman 2015). Both cases also emphasize that the relevant authorities in question may also take the form of Butlerian petty sovereigns in addition to titular executives. In short, logic of preemptive security affirms that, as Marieke de Goede puts it, "the unpredictable nature of the threat is not regarded as a reason for inaction"; as such, a preemptive approach "severs the relationship between the act of violence and the moment of security intervention, render[ing] action possible on the basis of suspicion alone" (2011, 506).

When considered from a juridical perspective, the implication of these points is that the adoption of a preemptive rationality entails a radical "transformation of [the] evidentiary bases for security action" which moves away from a grounding in verifiable knowledge and empirical fact and toward the juridically unbound realm of the speculative, the conjectural, and the purely imaginary (Aradau and van Munster 2011; de Goede 2011, 509). It is in this respect that the logic of preemptive security brings into being the first pillar of political exceptionalism—the suspension of the juridical order—since it suggests that a preemptive security rationality operates beyond the established rules of evidence that form the core of such an order. Indeed, when both the offense itself and the culpability of the offender exist only in a potential future imagined by the relevant authorities, independent legal mechanisms for adjudicating guilt and innocence are rendered ineffective since there is no clear way to explicitly demonstrate innocence once the accusatory imagination has been activated. In other words, the collection of empirically verifiable evidence and its evaluation according to the prescribed procedures which define any juridical order is simply not possible when dealing with an imagined future event, since the necessary reliance upon speculative and conjectural knowledge inevitably lowers to the point of irrelevance any evidentiary threshold that would circumscribe the sovereign prerogative to act upon a particular individual or group imagined to be potentially threatening (see de Goede 2011, 501, 508). Accordingly, the rules of evidence and procedure that form the foundation of a juridical order—particularly in the case of liberal democracies—are rendered incongruous with the circumstances, as an individual's demonstrable present innocence is immaterial when security threats are understood in terms of potential future acts that must be preempted. Questions of threat and culpability are thus instead placed under the direct purview of the relevant sovereign authorities, since they alone hold the prerogative to construct the imagined futures upon which such determinations must be based (Aradau and van Munster 2007, 106; Bonelli and Ragazzi 2014, 479; Ericson 2008, 67).

Considered in these terms, it becomes apparent that preemptive security rationalities bring into being a condition in which the established norms of the liberal juridical order not merely *do not* but in fact *cannot* apply, since they are simply incapable of accommodating the extended temporal horizons and attendant evidentiary uncertainties that are attached to questions of culpability for potential future acts that may or may not ever take place (Aradau and van Munster 2009, 697). Indeed, because it must "hold persons responsible for hypothetical deeds and their future effects," the logic of preemption "challenges law's temporal orientation toward past events" to the point of it "not being recognizable as law at all" (Opitz and Tellmann 2015, 119, 122). Put more simply, anticipatory security regimes suspend the juridical order *by default*, as their speculative epistemic foundations—which, again, are a direct function of the normative prioritization of the future—means that their constitutive practices will operate beyond the prevailing strictures of a (liberal) juridical order, particularly with respect to rules of evidence and procedure.

In regards to the connection between preemption and exceptionalism, these considerations are instructive, since the point is not merely that arbitrary preemptive interventions are *enabled* by the broader declaration of a state of exception vis-à-vis the specter of terrorism—a point that has been made at length in the existing literature. Rather, the present argument goes further by asserting that the logic of preemptive security *necessarily* brings about the suspension of the juridical order *at the very moment of its actualization*. In other words, the sorts of practices required to render its future-controlling normative imperatives actionable *must* operate beyond any legal framework premised upon the rules of evidence and process associated with conventional liberal judicial norms, since such practices are based upon an epistemic foundation rooted primarily in imaginative speculation rather than verifiable fact. Thus, wherever and whenever a preemptive security intervention takes place, the juridical order is suspended at precisely that point, since the very act of intervening in accordance with the logic of preemption "undermines the protection offered by liberal public law" by contravening the evidentiary and procedural norms through which it is constituted (Opitz and Tellmann 2015, 108). In this sense, preemptive security logically presupposes—and thus actualizes in practice—the first of the two pillars of political exceptionalism.

Practical Manifestations: The Rise of "Counter Law"

A useful illustration that highlights the above points can be found in the degree to which a number of liberal democratic states have resorted to what noted criminologist Richard Ericson has called "counter law" in the course of their adoption of preemptive approaches to (in)security governance (Ericson 2008; see also Opitz and Tellmann 2015, 117–22). As described by Ericson, "counter law" refers to the pervasive practice of creating what he terms "law against law" as a means of conferring the illusion of legitimacy upon erstwhile extrajudicial practices. Importantly, Ericson contends that counter law is deployed primarily to legitimate anticipatory security measures in the post-9/11 era, advancing a similar line of argument to that developed here by claiming that "preemptive security requires a radical reconfiguration of law" (2008, 57). Indeed, according to Ericson, under preemptive security rationalities, "new laws are enacted and new uses of existing law are invented to *erode or eliminate traditional principles, standards and procedures* of criminal law that get in the way of pre-empting imagined sources of harm" (57, emphasis added). This suggests that these "laws against law" are created as a response to precisely the conceptual issues discussed above, since the aim is to retain the veneer of juridical legitimacy while circumventing the prevailing evidentiary and procedural norms with which preemptive security practices are incompatible.

The examples offered by Ericson in support of his argument—primarily relating to the authorization of various preemptive measures by the USA PATRIOT Act (Ericson 2008, 62–68)—are by now somewhat dated. However, instances of "counter law" efforts to confer juridical legitimacy upon anticipatory security practices remain legion, as illustrated by several recent developments in both international and domestic jurisprudence related to the governance of terrorism. For instance, with respect to the international legal order, Tanja Aalberts and Wouter Werner document how changes in public international law are "mobilising uncertainty" in the context of the War on Terror to "create institutional mechanisms that put states under an obligation to adapt their legal systems in order to prevent possible terrorist activities" (2011, 2185). Perhaps the most prominent example here is the passage by the UN Security Council of Resolution 1373 in the immediate aftermath of the 9/11 attacks (United Nations Security Council 2001). Characterized as the "first-ever legislative act of the Security Council," Resolution 1373 uses the Council's Article 25 authority to effectively "globalise US counter-terror legislation"—and its preemptive elements in particular—by requiring UN member states to "ensure that their domestic legal systems are capable of preventing the commission and financing of terrorist activities" (Aalberts and Werner 2011, 2193). By making governments "responsible for *taking precautions* so that their country is free from *any possible future* terrorist activity," Resolution 1373 "disciplin[es] the legislative and executive machinery" of member states by requiring them to modify their legal orders to further legitimate the types of practices required for the anticipatory governance of terrorism (2194, emphasis added). That such a resolution was deemed necessary at all suggests that the types of practices required to comply with a preemptive global anti-terror regime are likely to violate existing juridical orders, and only a "counter law" strategy that embeds erstwhile exceptional practices into that order through radical modifications thereto can confer legal legitimacy upon them.

In a similar vein—and with a particular focus on liberal democracies—Filip Gelev has catalogued how the specter of a potentially catastrophic future embodied by the seemingly inevitable "next attack" has impelled judiciaries in a number of Western states to engage in what he terms "precautionary justice"—a practice that explicitly "*adapts existing legal concepts* to fit the requirements of risk society" by instituting new norms requiring "that decision-makers should take precautionary measures . . . where there is uncertainty about the nature and extent of the risk" (Gelev 2011, 2238–39, emphasis added). Importantly, this adaptive process has entailed "seeking ways to uphold radical laws" that "assess the present from the perspective of some possible future" and contravene the prevailing liberal legal order under which "courts adjudicate the guilt of persons *after* the violation of a legal norm" (2243–44, 2248, emphasis original). The practice discussed in chapter 3, wherein the American Federal Bureau of Investigation effectively manufactures terrorist plots that it subsequently preempts, with those implicated being later convicted through "preemptory prosecutions" that rely upon highly permissive readings of legal provisions against entrapment, offers a clear example of this aspect of counter law (Greenwald and Fishman 2015). Indeed, by inscribing "correlation rather than causality" into the juridical order, such practices legitimate key components of preemptive security in a way that typifies the "elimination of traditional principles" that Ericson identifies as the central element of "counter law" (Ericson 2008, 57).

In a third example, both international and domestic manifestations of counter law are visible in the emerging global regime for governing terrorist financing (de Goede 2012, 2011). Marieke de Goede's account of the preemptive character of the "blacklisting" regime that has been deployed by the European Union in this area clearly illustrates how the erstwhile extrajudicial practices associated therewith—such as anticipatory asset freezing and arbitrarily applied "targeted sanctions" that preclude the target from carrying out every day financial transactions—are becoming normalized through significant changes to the European legal order (de Goede 2011, 500; Sullivan and Hayes 2010). As de Goede puts it, these changes have entrenched "the use of political imagination within juridical action" by explicitly lowering the evidentiary thresholds for intervention and thus removing the decision to intervene from the established judicial mechanisms and vesting it in the person of the relevant sovereign decider (2011, 509). In this respect, the European Union's blacklisting regime operates in a similar manner to the Obama administration's anti-terror watchlists discussed in chapter 3 above. Indeed, the US Justice Department confirmed in court that decisions to add names to such lists—which precludes the individual in question from boarding a flight to or from the United States—are based upon "a standard of 'reasonable suspicion' that a blacklisted individual poses a threat." This threshold has been described as "a step below probable cause," which constitutes the established evidentiary standard for an arrest in American criminal law (Ackerman 2015). These are particularly instructive examples, since such reliance upon the imagination and the attendant contravention of established rules of evidence and due process are precisely the aspects of preemptive security that are at the core of its relation to exceptionalism.

It should also be noted that this rise of counter law has not been confined to the polities of the West. The Indian anti-terrorism regime—discussed briefly in chapter 3—is an instructive example here, as its legal foundations embody many of the same counter law characteristics as are found in other democratic polities. In this respect, the 2002 Prevention of Terrorist Activities Act (POTA)—which has been dubbed "India's Patriot Act" (Krishnan 2004)—and its successors, the 2005 Unlawful Activities Prevention Act (UAPA) and the 2008 National Investigation Agency Act (NIA), have all employed what amounts to a counter-law strategy to legitimate certain anticipatory anti-terror practices. For instance, supporters of POTA—which explicitly "treated terrorist acts outside normal criminal procedure" (Sasikumar 2010, 629)—marshalled counter-law arguments to defend it, often claiming "that terrorism can only be countered by quick and forceful action accompanied by *freedom from proceduralism*" (Mate and Naseemullah 2010, 264, emphasis added). Having been conceived in this spirit, POTA and its legislative successors embody a number of provisions that contravene legal norms conventionally associated with a liberal democratic polity in precisely the manner described by Ericson. One such feature is how these laws legitimate the preemptive detention of terror suspects by modifying the rules of evidence to the point of "introduc[ing] a new presumption of guilt in certain cases" (Mate and Naseemullah 2010, 175). Consider Section 20 of POTA, which "presume[d] that an individual charged with being a member of a terrorist organization is a terrorist unless that person can show that he or she has not participated in terrorist activities" (Krishnan 2004, 274). Section 49, meanwhile, permits anyone so charged to be detained without bail or trial for up to a year, but "[did] not require the prosecution to meet any evidentiary threshold when making its motion" toward this end (277). Such provisions invert perhaps the most vital norm of a liberal juridical order—the presumption of innocence—by placing the burden of evidentiary proof upon the accused rather than the state. This sort of fundamental rewriting of the rules of evidence and process is an archetypical example of counter law deployed in the service of preemptive security. It should be noted that POTA was allowed to lapse, and so these exact provisions are no longer in force; however, the UAPA and NIA have "resurrected significant portions of the old POTA regime," including the overarching emphasis on preemption and attendant modifications of evidentiary procedure. Regardless, the key point is that the counter-legal components of POTA were included for the express purpose of further legitimating a preemptive security regime, illustrating the conceptual point that such a regime's implementation in a democratic context requires a significant deviation from the juridical order.

Ultimately, the point is that both Ericson's theoretical description of "counter law" and the illustrative examples thereof just discussed highlight how the imperatives of preemptive security are effectively incompatible with the "principles, standards, and procedures" of a liberal legal paradigm. Indeed, the fact that governments intent on adopting preemptive approaches to terrorism must alter the prevailing juridical order in a manner that fundamentally undermines its spirit in order to maintain the veneer of legal legitimacy is particularly telling. It is no coincidence that Ericson explicitly embeds the concept of counter law—and thus, by association, the preemptive security practices it is used to legitimate—within the broader logic of political exceptionalism. He argues that "counter law is officially expressed as a state of exception," since it is enacted based on the belief that "normal legal principles, standards, and procedures must be suspended because of a state of emergency, extreme uncertainty, or threat to security with catastrophic potential" (Ericson 2008, 59).

The above examples are quite illustrative of this point as well, since each case suggests that the enaction of the anticipatory policies in question both was facilitated by and further entrenched a climate of political exceptionalism characterized by "imagery of radically uncertain yet potentially catastrophic threats" associated with the spectre of transnational terrorism (Aalberts and Werner 2011, 2184). For instance, de Goede (2011, 501) pithily argues that "blacklisting is an exceptional and preemptive measure *par excellence*"; Aalberts and Werner (2011, 2193–94) contend that "the political imaginary of uncertain yet undeniable catastrophe has informed the shift to a precautionary logic" in international law, which has manifested in the counter-legal requirements associated with UNSC Resolution 1373; while Gelev (2011, 2240–41) asserts that "because governments see any degree of likelihood of a catastrophic event as being too great to tolerate," an exceptional political climate in which "the judiciary adopts the logic of precaution" inevitably emerges, as in the case of the FBI actively manufacturing terrorist plots. With respect to the Indian context as well, Mate and Naseemullah explicitly argue that "discussions of POTA" are best understood "within a philosophical framework of 'states of exception'" and the "trend toward the 'executivisation' of law" that has characterized post-9/11 governance of terrorism worldwide (2010, 265).

The notion of counter law can therefore be understood as an attempt to disguise the exception as the norm, as it refers to a technique through which the violations of the juridical order that have been shown to be part and parcel of preemptive security are brought back into that order through significant modifications thereto. The crucial point, however, is that the degree of extra-judiciality requires these modifications to be so radical that the "principles, standards and procedures" upon which the juridical order is fundamentally premised are "erode[d] or eliminate[d]" (Ericson 2008, 57). In other words, by attempting to enfold the exceptional practices of preemptive security within the existing juridical order, counter-law strategies alter this order to the point where it is effectively indistinguishable from a purely exceptionalist political order. The somewhat ironic consequence is, therefore, that the enactment of counter law as a means of obfuscating exceptionalism in fact serves to further entrench it in practice, as the juridical order is altered to the point where it serves to actively facilitate the very exceptional practices that contravene its own constitutive norms. The phenomenon of counter law is a particularly useful illustration of the conceptual link between preemptive security and exceptionalism, and particularly the way in which the former brings into being a central component of the latter. Indeed, if any attempt to modify the juridical order in a manner that accommodates the imperatives of preemptive security must compromise the normative basis of that order, it follows that preemptive security practices themselves are always already in violation thereof and suspend it by default at the moment they take place. The radical modifications characteristic of counter law would not be required if this were not the case.

Put most bluntly, then, the very fact that a legislative strategy which by definition runs fundamentally "counter" to prevailing juridical orders has been widely deployed among liberal democracies to facilitate the adoption of preemptive security rationalities suggests that the latter can only be operationalized through practices that directly contravene the foundational legal norms of a democratic polity. It can be concluded that for reasons ultimately traceable to its normative prioritization of the future—since it is the reliance upon speculative knowledge required thereby that leads to the contravention of liberal legal norms—preemptive security rationalities logically imply the suspension of the juridical order, and because the latter constitutes one of two core components of political exceptionalism, this highlights the intimate conceptual link between preemption and exceptionalism by illustrating how the former presupposes the latter. This connection becomes even clearer, however, when considering how a politics of preemptive security also presupposes the enaction of a decisionist paradigm of sovereign authority.

Decisionism

Recall from the earlier discussion of exceptionalism that when the sovereign is freed from normative constraints through the suspension of the juridical order, a very particular form of political authority is enacted, whereby the sovereign no longer simply applies the law as created by a deliberative legislative body, but instead serves as "the *source* of law, as diktat" (Lazar 2006, 257). In other words, when the law no longer effectively limits the sovereign's prerogatives, it is the latter's will rather than any set of positive legal norms that constitutes the highest authority. The sovereign decision thus comes to possess what Derrida (1990) famously termed the "force of law," in that "acts that do not have the *value* of law"—such as unilateral proclamations by the executive—come to "acquire its *force*" by replacing officially legislated positive law as the primary legal foundation of the polity (Agamben 2005, 38, emphasis original). Such a paradigm of government reduces the praxis of societal governance to the normatively unconstrained decision of a sovereign authority, and thus is referred to as "decisionism."

Decisionism is the form of political authority that prevails within the proverbial state of exception, and constitutes the second "pillar" of political exceptionalism; however, its components are also analogous to the paradigm of sovereignty required to render preemptive security rationalities politically actionable. Indeed, just as the imperative to act on the future requires the epistemic primacy of the imagination—which, as we have seen, implies the suspension of the juridical order—the speculative basis of anticipatory action implies that the latter must be actualized through the type of pure sovereign decision associated with the theory of decisionism. The following discussion will expand upon and clarify this point by arguing that, for reasons that are again ultimately traceable to their underlying political temporalities, preemptive security rationalities logically presuppose what amounts to a decisionist paradigm of sovereign authority.

Before proceeding, however, it should once again be recalled that the use of the term "sovereignty" and its derivatives here refers to the expanded understanding described in chapter 3, and it can refer to both the executive branch of government and the myriad "petty sovereigns" on the front lines of contemporary (inter)national security strategies. Thus, while the theory of decisionism traditionally refers to the behavior of the executive, the present account is concerned with demonstrating how a politics of preemption enacts a condition in which the *characteristics* of decisionism obtain with respect to *any* sovereign decider. In other words, since my concern is with the logic of preemption in the abstract, the point is to highlight the purity of the preemptive decision as such, irrespective of the particular figure making it.

Conceptual Considerations

The core of this section's argument can be distilled down to the point that the imperative to govern the future through anticipatory interventions in the present logically requires what amounts to a decisionist paradigm of sovereign authority in order to be made actionable. This is because, as we have seen, a radically uncertain future can only be acted upon by way of effectively speculative decisions, and such decisions embody precisely the sort of pure singularity associated with the exercise of sovereign power under a decisionist framework.

To elaborate, recall that the political temporality of preemption requires decisions to be based primarily upon imaginative knowledge about potential futures divorced from any necessary link to empirical fact. While the preceding section showed how this places anticipatory interventions beyond the prevailing juridical order, this "epistemic primacy" of the imagination has important implications for the way sovereign power is exercised. In particular, it confers upon the sovereign decider a markedly enhanced degree of discretionary subjectivity that leads the anticipatory exercise of sovereign power to closely mirror that which is associated with theories of decisionism. This is because the futures against which preemptive decisions are framed possess no ontological presence apart from their existence in the realm of the imaginary; thus, any decision to intervene preemptively must be premised to a significant degree upon elements that are actively constructed by the sovereign decider (Aradau and van Munster 2009, 496–97). Thus, the imperative to act on the future augments the power of the sovereign to the point where it extends beyond merely weighing established evidence and making the final decision to act—which is the sovereign's defining prerogative—to also include constructing the primary epistemic basis for any interventionary decision through the exercise of the imagination. In other words, not only does the sovereign decide, but the preemptive decision necessarily takes place "under conditions of the sovereign's choosing" (Anderson 2007, 159). Thus, precisely because the uncertain future that is being acted upon constitutes something of an epistemic abyss, any sovereign decision to intervene preemptively "becomes in the true sense absolute," since only such a pure decision can provide the foundation for action in the face of the radical contingency that is being acted upon (Schmitt 2005, 12). The parallels to decisionist paradigms of sovereign authority are quite striking, as preemptive decisions embody precisely the sort of singular sovereign subjectivity called for by decisionist theorists such as Schmitt. It is also important to note that these characteristics of the preemptive decision apply, whether such a decision is made by a titular executive or a petty sovereign, since the purity of the decision is a function of the logic of preemption itself, and is enacted whenever any figure wields effectively sovereign power for anticipatory ends.

That a preemptive politics of security presupposes a decisionist paradigm of sovereignty should be unsurprising to those familiar with the work of Schmitt and other decisionists, since the idea that a powerful sovereign is necessary to adequately respond to the problem of irruptive contingency is, in fact, a key contention of the normative arguments put forth by such theorists. A brief consideration of these claims can serve to further illustrate the central point that preemption's underlying imperative to govern the temporal contingency of the future requires an augmented sovereign authority.

In this respect, theories of decisionism are premised upon a particular understanding of the human condition, in which the latter is viewed as inherently chaotic and violent, and thus defined by "the ever-present possibility of conflict" (Schwab 2005, xlii). This creates an originary political climate defined by "uncertainty and distress" that manifests as a "permanent potential emergency" whose "precise details . . . cannot be anticipated" (Lazar 2006, 258; Schmitt 2005, 6; Schwab 2005, xlii). For decisionists, governing this inherent contingency is the principal task of any political authority; however, the execution of this task will be severely hindered if the exercise of political power is constrained by the limiting strictures of a binding legal order to which the ultimate decider is subject. The putative problem with such a framework—which Schmitt refers to as "liberal constitutionalism"—is that it requires that all sovereign decisions be made in accordance with the prescriptive authorizations of a prevailing order of positive law (Schmitt 2005, 11). Decisionists view such limitations on sovereign authority as perniciously limiting, believing that the inherent unpredictability of what Schmitt pointedly calls "real life" ensures that no juridical order can provide a predetermined course of action for every possible contingency that may arise; as such, liberal constitutionalist polities are manifestly incapable of dealing with the "exceptional" irruptions with which they will inevitably be confronted (Schmitt 2005, 14; Schwab 2005, xliii). Put more simply, decisionists claim that because "[o]ne cannot tell in advance what circumstances might arise," responses thereto cannot be limited to "a matter of [the] simple application of a rule . . . dictated by norms" (Lazar 2006, 259–60).

Instead, decisionist theory sees the need for an ultimate authority to make a final decision that bridges "the gap between legal norms and facts" and allows action to be taken *vis-à-vis* the irruptive contingencies of "real life" that are "not codified in the existing legal order" (Huysmans 2008, 168; 2006, 148; Schmitt 2005, 6). For decisionists, only a juridically unconstrained sovereign endowed with the prerogative to make a unilaterally pure—and perhaps even arbitrary—decision is capable of fulfilling this requirement, since only such a figure is adequately liberated from the strictures of legal sanction to possess the decisional flexibility that is demanded. Accordingly, the logic of decisionism is ultimately premised upon enhancing the sovereign's discretionary power by "reintroduc[ing] a subjective competence . . . into a normative order that tends to focus on objectified processes" (Huysmans 2006, 149). This runs fundamentally counter to the liberal constitutionalist principle of the rule of law, which is "based on the rejection of all arbitrariness" and seeks to "repress the question of sovereignty by a division and mutual control of competences" (Lazar 2006, 261; Schmitt 2005, 11, 41). Decisionist theory holds that the unhindered sovereign discretion that liberal constitutionalism seeks to eliminate is, in fact, vital to the proper governance of an inherently uncertain, violently contingent world. In this sense, the decisionist argument is, perhaps somewhat ironically, an overtly normative one, since it is premised upon a very particular worldview in which the chaotically uncertain circumstances of the human condition simply cannot be adequately governed through an objective legalism, and instead require an absolute sovereign imbued with the prerogative to make a juridically unencumbered subjective decision (see Schmitt 2005, 55–56).

Returning to the relationship between decisionism and preemption, this *normative* argument at the core of decisionism closely parallels the *logical* argument about preemptive security developed in this section—namely, that the imperative to confront the uncertainty of the future *requires* what amounts to a decisionist paradigm of sovereignty in order to be made actionable. In other words, just as decisionism asserts the need for an unbound sovereign authority to "cross the distance between norms . . . and facts . . . to justify an intervention" (Huysmans 2006, 148), so preemptive security rationalities require such an authority to bridge the gap between radical uncertainty and anticipatory action, since this can only be accomplished through the leap of a pure decision in the face of an unknowable future. In this sense, the claim being advanced here that preemptive security presupposes decisionism is, in fact, prefigured by the claims made by decisionist theory itself, as a preemptive decision takes precisely the form of a Schmittian "decision in absolute purity" that is "created out of nothingness" (Schmitt 2005, 12–13, 66). The difference is that the decisionist claims are decidedly normative in nature—being rooted in a certain first principles assumption about the contingent nature of human life (Lazar 2006, 257–58)—while the claims being advanced here are logically extrapolated from the constitutive premises of preemption as a political rationality.

Thus, in much the same way as it suspends the juridical order by default, preemptive security enacts a decisionist paradigm of sovereignty out of what amounts to logical necessity, since its underlying imperative to govern the future can only be made actionable through the decisions of a sovereign authority exercising the form of quasi-absolutist power posited by decisionists as essential to any well-governed polity. Again, this is because the ultimate unknowability of the future demands that the epistemic basis for anticipatory decision be actively constructed by the decider, which in turn confers upon the sovereign an enhanced degree of discretionary subjectivity. The logic of preemptive security thus closely mirrors the logic of decisionist political theory in terms of the character of the decision—which in both cases must be "free[d] from all normative ties and . . . in the true sense absolute" (Schmitt 2005, 12)—and the underlying political aim—which is to confront and govern the "permanent potential emergency" wrought by the specter of contingency in human affairs (Lazar 2006, 258). And because a decisionist paradigm of sovereignty constitutes the second pillar of political exceptionalism, the conceptual connection between preemption and decisionism outlined here further supports this chapter's broader argument that the adoption of preemptive approaches to (in)security governance imply the enaction of an exceptionalist politics.

Practical Manifestations: The "Unitary Executive" and Indefinite Detention

It is again useful to flesh out these lines of theorizing by considering examples from the prosecution of the War on Terror, as there are a number of cases—particularly in the Western policy response to the threat of transnational terrorism—that show how preemptive security rationalities are operationalized through a decisionist paradigm of sovereignty. Perhaps the most conspicuous example in this regard was the George W. Bush administration's development and promotion of the so-called unitary executive doctrine. This distinctive approach to executive authority became the linchpin of the administration's counterterrorism program that was itself explicitly premised upon the logic of preemption (Ehrenberg et al. 2010). It also represented a prototypical manifestation of decisionist-type governance applied to the security realm. Accordingly, it provides an instructive practical illustration of the fusion between preemptive security and decisionism that has so far only been discussed in the abstract.

To elaborate, the central claim underpinning the doctrine of the "unitary executive" as formulated by the Bush administration was that "the executive's 'inherent authority' to protect national security permitted it to ignore laws that constrain the president" with respect to the carrying out of this core duty of office (Ehrenberg et al. 2010, 520). The powers of the president were construed to be such that the executive branch could choose to disregard the extant legal order created by the legislative branch if it deemed a particular statute to unduly circumscribe the president's scope of decisional authority with respect to national security questions. This articulation of executive authority places the executive branch above the law "as long as there [is] any connection to national security"—a caveat that can be (and, in practice, has been) read so broadly as to render the executive largely unbeholden to any juridico-normative limitations on its decisional prerogatives (526). A 2003 report prepared by the US Defense Department on the legality of the use of torture in the course of detainee interrogations illustrates the unitary executive argument clearly, asserting that "[i]n light of the president's *complete authority* over the conduct of war without a clear statement otherwise, criminal statutes *are not read as infringing on the president's ultimate authority* in this area" (quoted in Ehrenberg et al., 527, emphasis added). This statement is telling, as it almost casually asserts the unchecked authority of the executive when it comes to questions of national security—the determination of which, crucially, is *also* the prerogative of the executive. The national security-related prerogatives of the American president are here construed in a manner closely approximating those of the decisionist sovereign. Indeed, Schmitt's sovereign "decides whether there is an extreme emergency as well as what must be done to eliminate it," while "no higher authority [can] review the decision" (Schmitt 2005, 7, 56). This is precisely the paradigm of sovereignty embodied in the unitary executive theory, wherein the president-as-sovereign can ostensibly disregard the law under certain circumstances, but is also vested with the authority to decide when such circumstances obtain.

The decisionist character of the unitary executive doctrine is further evidenced by the way it was institutionalized in large part through President Bush's frequent use of so-called signing statements, which are declarations issued by the president when signing congressional legislation into law that affirms the executive's right to disregard that legislation should it be deemed to unduly constrain the president in carrying out his duties as commander-in-chief of the armed forces. Bush's unprecedented use of this process was highlighted in a report by the American Bar Association (ABA) Task Force on Presidential Signing Statements, whose key finding was that from the founding of the republic until the year 2000, sitting presidents used signing statements to challenge the laws they ratified fewer than six hundred times; yet Bush alone did so over eight hundred times in the first six years of his administration (American Bar Association 2006). In these addenda to congressional legislation, Bush would routinely affirm the decisionist character of the executive's national security authority by declaring that the law in question would be "construed in a manner consistent with the president's constitutional authority to supervise the unitary executive branch" (Bush 2006). While couched in somewhat sanitized language, the implication of this statement—which was often repeated verbatim in each discrete signing statement—is that the new law's provisions could be disregarded by the executive branch should the latter unilaterally deem it a hindrance to the prosecution of the War on Terror, thus suggesting that "even the Supreme Court did not have the last word on legal rulings" relating to national security questions (Ehrenberg et al. 2010, 514). The ABA's conclusion that such practices were "contrary to the rule of law and our constitutional system of separation of powers" emphasizes the degree to which the Bush administration's performative enaction of a unitary executive represented a practical manifestation of decisionism (and thus exceptionalism) (American Bar Association 2006). Indeed, it is precisely the sorts of normative constraints contained in the statutes to which signing statements were appended that are decried as excessively limiting in decisionist theory.

Moreover, the unitary executive doctrine was often explicitly defended in terms quite similar to normative decisionist arguments relating to the proper governance of a dangerously contingent world. Proponents frequently argued that "new threats to domestic security required an extraordinary concentration of executive power," since "only unrestrained presidential power could wage the 'War on Terror' and protect citizens at the same time" (Ehrenberg et al. 2010, 512, 514). This is an important point, because it not only further emphasizes that the unitary executive doctrine constitutes a clear manifestation of decisionism but also highlights a connection to the logic of preemption. Indeed, the adoption of this decisionist paradigm of executive power can be seen as a direct outgrowth of the Bush administration's broader post-9/11 security agenda, whose defining feature was its commitment to the logic of preemption.[9] This is evidenced by the fact that administration officials regularly defended the concept of a unitary executive by invoking the overarching imperative to act preemptively against the possibility of future attacks—for example, by referencing "the president's recognized authority . . . to *prevent further catastrophic attacks* on the homeland" to advance the claim that this could not be accomplished without a sweeping reformulation of executive power along what amounted to decisionist lines (United States Department of Justice 2006, 2, emphasis added; see also Ehrenberg et al. 2010, 512–14).

The Bush administration's embrace of the unitary executive doctrine can be understood as a direct function of its commitment to the broader imperative to preempt any potential future irruption of terrorist violence. As Ehrenberg et al. put it, "The administration's unitary vision of an executive-dominated government . . . was organically linked to its belief in . . . the preemptive" approach to national security (Ehrenberg et al. 2010, 516). Moreover, this connection between preemptive security and an effectively unconstrained executive authority has also seemingly been embraced by the Obama administration. Again, the example of anti-terror watchlists is instructive, as the addition of names to such lists has been described by government officials as being based on "predictive assessments of potential threats." And crucially, while defending this practice in a federal court in May 2015, two Justice Department officials directly asserted that "[j]udgments concerning such *potential threats* to aviation and national security *call upon the unique prerogatives of the executive* in assessing such threats" (quoted in Ackerman 2015, emphasis added). The clear implication is that only the executive is adequately equipped to successfully identify and act preemptively against potentially catastrophic threats, and should not be constrained by the judiciary from exercising this capacity to the fullest in carrying out its role as guarantor of national security. Continuity with the Bush-era conception of a unitary executive is quite apparent.

Ultimately, by developing the unitary executive theory in accordance with the logic of decisionism, and defending its necessity with reference to the logic of preemption, it is evident that post-9/11, US national security officials saw the adoption of what amounted to a decisionist paradigm of sovereignty as the necessary means through which to realize the end of creating a functioning preemptive security regime. Such a conclusion is supported by Ehrenberg et al., who, near the end of their comprehensively documented account of Bush-era national security policy, pithily sum up the latter as defined by "a doctrine of preemption abroad coupled with a theory of a 'unitary executive' at home" (Ehrenberg et al. 2010, 512). This "coupling" is important, since it suggests that a preemptive approach to terrorism and a decisionist approach to sovereign authority were considered by the Bush administration to be two sides of the same coin—a line of reasoning that continues to be evidenced by the Obama administration's anti-terror policies as well. The American embrace of a unitary executive doctrine thus provides a useful illustration of precisely the fusion of preemptive security and decisionism—and, by extension, exceptionalism—suggested by the conceptual arguments developed in this chapter.

Given its decisionistic character, the adoption of the unitary executive doctrine predictably led to a number of overtly "exceptional" counterterrorism policies that operated beyond the prevailing juridical order. One such policy—the practice of indefinitely detaining individuals merely suspected of engaging, or even planning to engage, in terrorism-related activities—provides an even more specific illustration of the link between the preemptive imperative and a decisionist paradigm of sovereignty than the more general notion of the unitary executive, and thus merits brief consideration for the purpose of further fleshing out this conceptual connection. The key point here is that indefinitely detaining suspected terrorists constitutes an exemplary exercise of preemptive security, while the form of sovereign power through which it operates is archetypically decisionist in nature for reasons directly related to this anticipatory character.

The practice of indefinite detention that emerged in the early days of the War on Terror refers to the incarceration of individuals whom the relevant authorities—which, in this case in particular, often constitute prototypical "petty sovereigns"—"fear *could* pose a danger of terrorism *even if they have little evidence*" (Butler 2006, 68, emphasis added). Indefinite detention "criminalizes suspects for *imagined future harm they might cause*, rather than past crime," and constitutes an anticipatory intervention congruent with the preemptive imperative to govern the potentially catastrophic future portended by the spectre of the next terrorist attack (Ericson 2008, 63, emphasis added). We have seen that such an emphasis on the potential requires that the evidentiary threshold for any interventionary decision be diminished significantly (de Goede 2011), and the decision to detain indefinitely is no exception. As Butler pithily puts it, "Conjecture is the basis of an indefinite detention" (Butler 2006, 69). We have also seen that such a conjectural basis for decision-making enhances the arbitrary discretion of the sovereign decider, since the decision must be based upon an epistemic foundation that is primarily of the decider's own imaginative construction. The explicitly preemptive practice of indefinite detention functions through what amounts to a decisionist paradigm of sovereignty, since the inherently speculative basis of the decision to detain requires that the decider possess a degree of unilateral discretion reminiscent of that ascribed to the Schmittian sovereign.

Butler captures this point well in her argument about the centrality of the process of "deeming" to the indefinite detention regime developed by the US government in pursuit of its broader preemptive security strategy (Butler 2006; see also de Goede 2012). In this respect, Butler emphasizes the degree to which indefinite detention is premised upon the exercise of pure decisional discretion, arguing that "the decision to detain someone indefinitely is a unilateral judgment made by government officials who simply deem that a given individual or, indeed, group poses a danger to the state" (Butler 2006, 58–59).[10] In other words, under a security regime whose underlying aim is to preempt future irruptions of terroristic violence, "the 'deeming' of someone as dangerous is sufficient to make that person dangerous and to justify his indefinite detention" (59). Consequently, decisions to detain take place beyond any juridical restrictions on the decider's discretionary prerogatives, as they "are not decisions . . . made by a judge, for which evidence must be submitted in the form of a case conforming to certain protocols" (76). Rather, the act of "deeming someone dangerous is an unsubstantiated judgment that . . . works to preempt determinations for which evidence is required," conferring upon the decider a "license to brand and categorize and detain on the basis of suspicion alone" (76). In its capacity as a preemptive security intervention, therefore, the practice of indefinite detention "dispenses with conventional evidentiary procedure" by requiring a sovereign authority that has "the final say in matters of guilt and punishment" in lieu of the established judicial system (75, 71). Such an authority thus "assumes a lawless and yet fully effective form of power" that quite closely mirrors that of the Schmittian decisionist sovereign (59). The practice of indefinite detention thus constitutes an archetypical manifestation of preemptive security—since it is concerned with actively governing an imagined future threat—and yet also constitutes an archetypical manifestation of the decisionist paradigm of sovereign authority—since it relies upon the arbitrary process of "deeming" to be rendered operational. When considered in relation to the conceptual arguments developed above, it should be clear that this is no coincidence, and that in fact its decisionist nature is a direct function of its preemptive character.

Fleshing out this latter idea has been the aim of this section, as it is key to understanding the broader symbiotic relationship between preemption and exceptionalism being posited in this chapter. I have argued that the type of decision required to render a preemptive security imperative actionable can only be made in what amounts to a decisionist manner, as the ultimate unknowability of the future against which it seeks to act requires that the sovereign decider be granted radically enhanced discretionary powers tantamount to those advocated by thinkers like Schmitt. The Bush administration's articulation of the "unitary executive" doctrine as a central component of its preemptive approach to governing terrorism lends empirical support to these points, as does the way in which the anticipatory practice of indefinite detention operates through the conjectural decisional process of "deeming." Finally, it should once again be recognized that this link between preemptive security and decisionism is ultimately rooted in the political temporality of preemption, as it is the underlying imperative to govern the future through anticipatory action in the present that creates the need for the sort of speculative interventions that can only be actualized through what amounts to a decisionist paradigm of sovereign authority.

CONCLUSION

A principal aim of this book is to think through what is at stake with the widespread proliferation of anticipatory governance practices in general by developing an in-depth theorization of a particularly conspicuous manifestation of this trend in the realm of global (in)security governance. Toward this end thus far, chapter 3 offered an account of what "preemptive security" can be understood to mean in the contemporary context, while chapter 4 provided an abstract account of how the logic of preemption functions as a political rationality. This fifth chapter has sought to narrow the analysis by developing a more detailed account of the sort of politics that is made possible—or, perhaps more accurately, logically required—by such a rationality in the security context. Most basically, it was argued that by placing speculative knowledge at the core of the sovereign decision, the future-oriented political temporality embedded in the logic of preemption requires that security rationalities based thereupon be actualized through an exceptionalist political paradigm characterized by a suspended juridical order and a decisionist form of sovereignty. In addition to helping flesh out the idea of preemptive security on a theoretical level, these arguments also add to the existing literature in critical security studies by providing a conceptual explanation for the widely chronicled empirical correlation between the rise of preemptive security strategies and the proliferation of political exceptionalism in the context of the War on Terror. Moreover, by demonstrating how the relationship between preemption and exceptionalism is ultimately traceable to the former's political temporality, the above arguments further highlight the utility of employing a deliberately "temporal lens" for political analysis. Indeed, it is important to recognize that the critical insights developed in this chapter only became possible by foregrounding questions of temporality and futurity, thus supporting the wider methodological point discussed in the introduction. The next chapter will continue to foreground such questions as it builds upon the preceding arguments by considering the implications of adopting preemptive security rationalities for the experience of political subjectivity, particularly in the liberal democratic societies that have been at the forefront of this trend.

NOTES

1. It should be noted that almost one year after this speech, in February 2013, a Justice Department White Paper containing a more formal articulation of the Obama administration's legal rationale for the targeted killing program was leaked to the press (see United States Department of Justice 2013). While addressing a number of key questions in somewhat more detail than Holder's speech, the underlying substance of the White Paper's argument is largely identical to that of his 2012 remarks. Thus, while it may not be the most recent or detailed articulation of the administration's targeted killing policy, I use the text of Holder's speech as the basis for the introduction to this chapter, since it accurately represents the Obama administration's stance on the question of targeted killing; in addition, its greater concision offers a clearer illustration than the White Paper of the centrality of the logics of preemption and exceptionalism to this stance. It should also be noted that all subsequent quotations from the speech are derived from the source cited in the first sentence, and thus will not be explicitly referenced for the remainder of this introductory section.

2. For example, during the 2012 presidential campaign, the Obama reelection team repeatedly emphasized the president's "strength" on foreign policy and national security by invoking the apparent decimation of al-Qaeda via the drone program deployed across the greater Middle East.

3. Note how the language directly alludes to the reasoning advanced in the 2002 *National Security Strategy* in which, as discussed previously, the logic of preemption was first explicitly elaborated as the basis for the conduct of the War on Terror (see Ehrenberg et al. 2010, esp. chapter 10).

4. On this point, see works by Derrida (2003), Jarvis (2009), and Lundborg (2012).

5. At this point, a brief terminological explanation is necessary. In addition to imagination, the existing literature on the politics of preemption uses several terms used to describe its modes of knowledge creation, including "speculation"—as used by Marieke de Goede in her work on the governance of terrorist financing (de Goede 2012)—and "conjecture"—as discussed by Aradau and van Munster in their study of the contemporary "politics of catastrophe" (Aradau and van Munster 2011). The latter work also provides a useful distinction between purely speculative knowledge and knowledge derived from what the authors label, following Carlo Ginzburg, "conjectural reasoning"—which specifically refers to the use of fragmentary evidence and somewhat arbitrarily determined "clues" to develop "possibilistic" accounts of the future upon which to act in the present (50–51; see also Bonelli and Ragazzi 2014, 478). While certainly valuable, I am not concerned with such taxonomical distinctions in this discussion. This is because the aim is merely to emphasize how the logic of preemption operates primarily through the use of "knowledge" about an ultimately unknowable future, which is thus divorced from the strictures of empirical facticity. Thus, while I primarily use "imagination" and its derivatives, terms such as speculation, conjecture, and their own derivatives are also used largely interchangeably, since this key characteristic is captured by all three terms and transcends their particularistic definitional distinctions.

6. While this point about the epistemic primacy of the imagination has been widely recognized in the critical literature on preemptive security, it should be noted that Louise Amoore has productively complicated this idea somewhat. In a recent article, Amoore

compellingly argues that anticipatory security strategies do not eschew scientific calculation in favor of speculation and conjecture, arguing instead that "calculability is never in question" and preemptive security's quotidian mechanisms still operate through calculative rationalities that more explicitly incorporate intuitive reasoning (Amoore 2014b, 425). In this sense, the intuitive dimension of preemptive security does not so much supplant rational calculation as become embedded within it, and, through a fascinating engagement with a 1939 debate between Wittgenstein and a young Alan Turing, Amoore claims that this in itself is nothing new, and in fact represents merely the latest iteration of "an historic enfolding of the intuitive faculties within mathematical calculation" (425). The upshot is that contemporary preemptive security is concerned primarily with "the establishment of routines, subroutines, and procedures" based around these modified calculative rationalities, and is thus largely "not imaginative at all" (427). Because Amoore's intervention disturbs the prevailing consensus and provides compelling insights into the specific ways imagination is mobilized in the practical operationalization of preemptive security, it cannot be ignored in the context of this chapter; however, her arguments do not—and are seemingly not intended to—wholly refute the broader conceptual point relating to the centrality of the imagination to the *logic* of preemption as a political rationality, which is the primary focus of the present discussion.

7. The discussion of Attorney General Holder's remarks that opened this chapter suggests that this is precisely the approach taken by the Obama administration with respect to the relationship between the rule of law and executive authority in the context of terrorism governance, as the existential threat purportedly posed by transnational terrorism was deemed by Holder to not only permit but also necessitate such "exceptional" exercises of sovereign power as the targeted killing of a US citizen by his own government.

8. Indeed, exceptionalism refers to the diminution of limits upon the exercise of what can be best understood as sovereign power, and is conventionally associated with an empowered executive. However, in accordance with the Butlerian understanding of sovereignty discussed near the end of chapter 3, it should be emphasized that it can also manifest through the behavior of other actors capable of exercising sovereignty. In other words, it is important to recognize that because executives and "petty sovereigns" can exercise what amounts to "sovereign" power, both are equally capable of serving as conduits for a politics of exceptionalism in practice.

9. This latter point is unambiguously illustrated by the wording of the 2002 *National Security Strategy* that has been cited several times elsewhere in this book. In laying out the governing rationality for the nascent War on Terror, the document forcefully asserted that "the United States can no longer rely on a reactive posture"; as such, "even if uncertainty remains as to the time and place of the enemy's attack . . . the United States will, if necessary, act preemptively" (White House 2002).

10. Again, we see the importance of the petty sovereign here, as it is these "government officials" who are delegated with what amounts to sovereign power in their capacity to make the sorts of preemptive decisions upon which indefinite detention practices are based. In fact, it is in specific relation to the preemptive practice of indefinite detention that Butler formulates her conception of the petty sovereign in the first place. See the discussion of the concept of sovereignty in chapter 3 above.

Chapter Six

Precarious Subjectivities, Drone Warfare, and Autoimmunity

The preceding chapter explored how the political temporality embedded in the logic of preemption requires that anticipatory security rationalities be operationalized through what amounts to a politics of exceptionalism. This chapter builds upon that argument by further considering what this implies for the practical implications of adopting preemptive approaches to (in)security governance. In other words, having thus far considered what preemption "is" conceptually, how it "works" as a political rationality, and the paradigm of sovereign authority that it requires to be implemented as a security strategy, this chapter is concerned with thinking through the sorts of subjective conditions that are thereby created. The discussion will examine what the relationship between preemption and exceptionalism, described in the previous chapter, implies for the character of political subjectivity under a preemptive security regime. In particular, it will be argued that the exceptional paradigm of sovereignty presupposed by the logic of preemptive security effectively eliminates any juridico-normative mediation between sovereign and subject. This renders the latter perpetually vulnerable to unilaterally decided anticipatory interventions by the former—a condition that can be best described as precarious.

The first section of the chapter will consider this claim in greater conceptual detail by further unpacking the relationship between the sorts of decisionist sovereign figures required by preemptive security rationalities and the political subjects over whom they exercise authority. It will be contended that this relationship conspicuously resembles that which obtains in the "exceptional spaces" theorized extensively by Agamben and other theorists of the state of exception (2005, 1998). Following a similar pattern to the previous chapter, the second section will flesh out these conceptual claims by discussing them in terms of a particular illustrative example drawn from the ongoing War on Terror—namely, the targeted killing of Anwar al-Awlaki that was briefly discussed at the beginning of chapter 5. The al-Awlaki case is instructive because it represents an archetypal manifestation of the sorts of practices that are necessarily made possible under a preemptive security regime. Indeed, the extrajudicial killing of an American citizen by the US government on foreign soil illustrates precisely the sort of precarious subjectivity that is part and parcel of a preemptive approach to (in)security governance.

In light of this discussion, the third section takes a step back and considers what the preceding arguments suggest about the coherence of preemptive security as an idea. It is argued that its coherence can be called into serious question, since the precarious subjectivities it enacts closely resemble the types of conditions that it is normatively premised upon diminishing. In other words, I contend that the arguments developed in the first two sections of the chapter—and in part II of this study more broadly—indicate that preemptive security ultimately ends up bringing into being that which it is premised upon eliminating. Accordingly, the fourth and final section of this chapter considers how the politics of preemptive security can be productively read in terms of Derrida's conception of "autoimmunity" (Derrida 2003, 2005), which usefully captures the sorts of intrinsic conceptual tensions that have been identified. The chapter concludes by suggesting that, in light of these claims, both the practical policy validity and the abstract conceptual coherence of preemption as a political rationality can be seriously challenged—the broader implications of which are considered in the final chapter.

PREEMPTIVE SECURITY AND/AS PRECARIOUS SUBJECTIVITY

To help understand the wider implications of the proliferation of preemptive security rationalities, it is useful to consider the effects this has on the character of political subjectivity, particularly in the context of the liberal democracies at the forefront of this trend. A key point in this respect is that the exceptionalist paradigm of sovereign authority that a preemptive security politics requires to be practically operationalized brings into being a relationship between sovereign and subject in which the latter is rendered perpetually vulnerable to the sudden irruption of arbitrary and possibly violent interventions by the former. That is, the condition of political subjectivity implied by the logic of preemptive security is decidedly precarious. To be sure, such claims may at first glance seem excessively alarmist, and perhaps even hyperbolic. Yet a degree of extrapolative reflection upon the arguments developed in the preceding chapters suggests that the creation of such precarious subjectivities is a logical consequence of adopting preemptive approaches to (in)security governance.

In this respect, recall once more that the logic of preemption's concern with governing the future confers "epistemic primacy" upon the imagination (Aradau and van Munster 2011, 85). We have seen that this has the effect of radically enhancing the discretionary capacity of the relevant sovereign authority—whether titular executive or "petty" functionary—since the latter must actively construct the imagined potentialities that form the primary epistemic basis for anticipatory action. We have also seen that because the future being acted upon remains inherently unknowable, it follows that in the course of imagining possible futures against which to act in the present, no potentiality can be *a priori* excluded from consideration. This stems from the explicitly precautionary aspect of the post-9/11 politics of preemption, which demands that decision-makers "take the most far-fetched forecasts seriously" because the very uncertainty they are seeking to tame precludes the summary dismissal of any imaginable possibility (Daase and Kessler 2007; Ewald 2002, 288; Opitz and Tellmann 2015, 112). The result is that a preemptive politics of security manifests as a regime in which, by default, "everyone is suspect" in the eyes of the sovereign—and it is under such conditions of "universal suspicion" that anticipatory action must be taken (Ericson 2008, 66). Moreover, the previous chapter has shown that such action operates beyond the juridical order and through a decisionist paradigm, meaning that normative limits on the scope of decisional possibility are all but eliminated. In other words, the only effective limitations on the exercise of sovereign power in this context are the limits of the sovereign imagination itself.

To put these points another way, under a politics of preemption, the sovereign is tasked with taming the future's radical contingency, and because the epistemic basis for any action toward this end consists of an imagined potentiality of the sovereign's own construction, individuals are either targeted or ruled out based on the subjective discretion of the sovereign rather than any objective juridico-normative guidelines. Since none can be *a priori* excluded, all are always already potential targets for anticipatory action. What emerges is a political condition in which *any* individual may be subjected to what amounts to a unilaterally decided intervention at any time, since the futurity of the threat means that no one can be summarily absolved from suspicion and it takes only the whim of the relevant authority to deem an individual to be a threat and thus precipitate an anticipatory response. This has the effect of rendering all subjects perpetually vulnerable to a potentially violent intervention based on the conjectural imagining of a future that may not ever come to pass—a decidedly precarious condition.

Now, it is worth noting that this abstract conceptual point risks some oversimplification—particularly when expressed in the rhetorically effective language employed, for example, by Ericson, as quoted above—since it could be read as implying a latent equality that is in practice quite absent from the post-9/11 politics of (in)security. In other words, the logic of preemption does logically imply a security regime in which *all* individuals are placed on a "continuum of risk," since none can be *a priori* absolved of posing a potential threat in an ultimately unknowable future (Ericson 2008, 66); yet the practical politics of preemption—which determines *where* a particular individual is placed on this continuum—will inevitably reflect the broader discourses of threat construction that underlie all forms of social sorting in the name of security. Thus, while perhaps "everyone" must indeed be suspect to at least some extent under a preemptive security regime, it is equally important to recognize that the degree to which a particular individual or group is treated as such by security apparatuses will vary quite widely depending upon a variety of politico-discursive factors exogenous to the logic of preemption itself. But this does not undermine the key conceptual point that a preemptive security rationality cannot in theory rule out any potential threat and thus renders subjects immanently vulnerable to irruptions of sovereign violence.

It is also true that this vulnerability is rarely translated into acts of arbitrary violence, and in fact is quite unlikely to ever do so for almost every individual living under a regime of preemptive security. Yet this empirical fact is largely immaterial to the broader conceptual point being made, because this ever-present *potential* for anticipatory sovereign intervention based upon speculative imaginings of potential futures still implies that the subjective condition enacted by a politics of preemptive security can be best described as precarious. Indeed, it is this permanent state of susceptibility to such interventions that characterizes the subjective experience, because even if no such action ever takes place, the *constant possibility* that it will is a defining feature of the prevailing relationship between sovereign and subject. This possibility is what follows from the enhanced decisional discretion that we have seen is vested in the sovereign by the logic of preemption, since anticipatory decisions are located outside the circumscriptions of the juridical order thus removing the normative barriers that protect individuals from being targeted on the basis of conjectural knowledge that they have no capacity to contest. The point, therefore, is that even if such arbitrary targeting never actually takes place—which will most likely be the case for almost everyone—this will not be because of any normative constraints upon the sovereign's decisional authority; rather, it will be because the sovereign happened to decide against it. In other words, even if the relevant authorities never exercise the decisionist prerogatives conferred upon them by the logic of preemption, the mere existence of these prerogatives signifies that the *potential* for arbitrarily violent irruptions of sovereign power remains ever-present. A perpetual exposure to such interventions defines the subjective condition under a preemptive security regime, characterizing this condition as inherently precarious.

Returning to the relationship between preemption and exceptionalism, these points further highlight this connection by demonstrating that the relations between sovereign and subject enacted by the logic of preemptive security closely resemble those that characterize the particular spaces in which the political exceptionalism theorized by Agamben via Schmitt is manifested. In his description of such a space—which he generically terms the "camp"—Agamben argues that sovereign power confronts its subjects "without any mediation," since the sovereign's ability to act is unbound by juridical norms while subjects are deprived of any agentic capacity to resist or contest its unilateral decisions (Agamben 1998, 171; 2000, 41). This is precisely the sort of relation that is created under a preemptive security regime, since on the one hand, the imperative to govern the future demands a sovereign authority that must decide to act on the basis of its own imagination and cannot be limited by the juridical order, while on the other, those targeted for anticipatory intervention by such a decision cannot contest or resist it, since any demonstrable present innocence is ultimately irrelevant when the basis for intervention is an imagined future that may never come to pass. Accordingly, just as in the proverbial "camp," the *possibility* that any individual may be arbitrarily subjected to potentially violent sovereign interventions is always there under a preemptive security regime. Indeed, even within the boundaries of sovereign states ostensibly committed to the rule of law and upholding of human rights norms, the adoption of a preemptive security rationality enacts a political condition characterized by the ever-present potential for anyone to be inscribed as the sort of "bare life" against which "everything is possible" described by Agamben (1998, 170). The subjective experience that is created by such a politics of potentiality is therefore best described as precarious.

THE KILLING OF ANWAR AL-AWLAKI

A helpful illustration of these points can be found in the incident briefly discussed at the opening of the preceding chapter—the targeted killing of suspected al-Qaeda operative and US citizen Anwar al-Awlaki, who, in September 2011, was killed by an American drone strike as his vehicle traveled down a rural road in northern Yemen. This case is particularly instructive because it clearly illuminates both the type of action on the part of sovereign authorities which is made possible by the logic of preemptive security and the concomitantly precarious subjectivities that are thereby enacted.

To elaborate upon these points, it must first be established that the Obama administration's decision to target al-Awlaki in fact constituted an explicit exercise in preemptive security. This is evidenced, on the one hand, by the point that the broader CIA "Killing Program" of which it is a high profile example has been framed in precisely such terms by its proponents (Kessler and Werner 2008; Leander 2011). For example, in a 2012 interview with the *New York Times*, former National Counterterrorism Center head Michael E. Leiter claimed that the practice of targeted killing via drone strike was embraced by President Obama as the most appropriate response to the "situation where he is being told people *might* attack the United States tomorrow" (quoted in Becker and Shane 2012, emphasis added). This suggests that Obama and his administration view drone warfare as an effective way to ensure that such a catastrophic future does not come to pass—that is, in explicitly preemptive terms. Moreover—and with respect to the al-Awlaki case in particular—the preemptive character of the killing is highlighted by the administration's *post hoc* framing of the incident. While acknowledging the juridically extraordinary character of the act, the administration sought to legitimate the killing after the fact by specifically invoking a preemptive imperative. In this respect, it was simply asserted that al-Awlaki "posed some sort of imminent threat"—the precise nature of which was not specified—which justified "extraordinary measures"—in this case, the killing of a citizen by his own government without what is conventionally understood as due process (Koring 2011). This suggests that the decision to target al-Awlaki was made on the basis of an imaginarily constructed future in which that ambiguous potential threat *had*, in fact, manifested as the next major terrorist attack on the United States. In relying upon an imagined future in this way, the targeting of al-Awlaki constituted an archetypical manifestation of a preemptive security rationality in action.

Besides being an exemplar of preemptive security, however, the targeted killing of al-Awlaki also constituted an archetypically "exceptional" act, as it exemplified both pillars of political exceptionalism described in the preceding chapter. With respect to the first—the suspension of the juridical order—the act was patently extrajudicial, and it suspended the legal order at the moment and point of its occurrence. Indeed, the killing was not authorized through conventional judicial channels or in accordance with prevailing standards of evidence; as such, it was both ordered and carried out absent the due process of law guaranteed to all US citizens under the Fifth Amendment to the Constitution. Moreover, that killing al-Awlaki by remote drone strike would contravene the juridical order appears to have been apparent to the Obama administration in that, much as the Bush administration sought to circumvent existing legal limitations on presidential authority by invoking the "unitary executive" doctrine, the Obama White House sought to justify the killing through a line of legal reasoning premised upon enhanced executive discretion with regard to issues of national security. This is evidenced by an internal memo from the Justice Department, prepared with specific reference to the al-Awlaki case, which asserted that "while the Fifth Amendment's guarantee of due process applied, it *could be satisfied by internal deliberations in the executive branch*" alone (quoted in Becker and Shane 2012, emphasis added).

This circumvention of the juridical order through appeals to executive privilege hints at how the al-Awlaki case also embodies the second pillar of exceptionalism—a decisionist paradigm of sovereign authority. In this respect, the decision to target al-Awlaki proceeded from the Justice Department's aforementioned position and took place in accordance with the Obama administration's approach to drone warfare that "concentrates power over the use of lethal U.S. force outside war zones within one small team at the White House" (Dozier 2012). This process was chronicled in detail by a *New York Times* investigative report published amid some controversy in May 2012, which is worth quoting at length to illustrate the degree to which the exercise of sovereign authority underpinning the drone campaign takes a decisionist form:

> Every week or so, more than 100 members of the government's sprawling national security apparatus gather, by secure video teleconference, to pore over terrorist suspects' biographies and recommend to the president who should be the next to die. This secret "nominations" process is an invention of the Obama administration, a grim debating society that vets the PowerPoint slides bearing the names, aliases, and life stories of suspected members of al Qaeda's branch in Yemen or its allies in Somalia's Shabab militia. . . . [N]ames go off the list if a suspect no longer appears to pose an imminent threat. . . . The nominations [then] go to the White House, where by his own insistence and guided by [chief counterterrorism advisor Jim] Brennan, Mr. Obama must approve any name. (Becker and Shane 2012)

This executive-centred approach is archetypically decisionist in nature, as the final authority to decide who is to be killed and when the strike is to take place is both excused from the limitations of the existing juridical order and explicitly granted to the president alone. Indeed, Obama's former chief of staff William Daley is quoted in the same article as stating that the president believes his status as commander in chief requires that "his own judgments should be brought to bear on strikes." Thus, operating under the presupposition that executive branch deliberation followed by the president's final decision constitutes adequate due process, the administration's preemptive use of drone warfare frees the executive branch from any normative circumscriptions regarding this use of violence, thus vesting within the person of the president the discretionary capacity to determine who to target, when to strike, and crucially, *what counts as adequate evidence* that someone poses a threat sufficient to merit being killed (Ackerman 2012). Such prerogatives conspicuously mirror those ascribed to the sovereign under the decisionist paradigm described in the preceding chapter, in that the president decides both when and where to intervene, and he must also conjecturally construct the ultimate epistemic/evidentiary foundation upon which to make this decision.

The al-Awlaki case offers a further illustration of the relationship between preemption and exceptionalism. However, with specific relevance to the concerns of this chapter, it also forcefully highlights the practical implications of this link for the character of political subjectivity under preemptive security regimes. Specifically, it highlights the degree to which the anticipatory exercise of sovereign power in this context shifts the relation between sovereign and subject in the direction of an unmediated confrontation. Indeed, in the al-Awlaki case, the nature of the sovereign/subject relation was such that once the president made the preemptive decision to target al-Awlaki, the latter could immediately be killed with impunity by the agents of American sovereignty. In other words, the law no longer served as a mediator between sovereign and subject, since despite being a US citizen, al-Awlaki could nonetheless be killed purely on the basis of what amounts to an executive decree. The juridico-normative limitations on the president's decisional authority in this respect were thus subordinated to a preemptive imperative with which such limits are fundamentally incompatible. Moreover, the decision also deprived al-Awlaki of the capacity to contest or resist this precarious status, as the extrajudicial character of the targeting process shields the resulting decisions from the degree of judicial review that the Fifth Amendment purportedly guarantees all citizens. Thus, faced with a sovereign power that could unilaterally decide when he would be killed and against which he had no immediate legal recourse, al-Awlaki was effectively inscribed as precisely the sort of "bare life" that Agamben and others associate with the "exceptional" space of the "camp" without actually being located in such a space (Agamben 1998, 2000).[1]

The al-Awlaki case provides a clear example of the sort of exceptional action that is made possible by the logic of preemptive security, while also demonstrating that such exceptional interventions take place at the discretionary will of the sovereign. In other words, it shows that, under the logic of preemption, the sort of violent, unilaterally decided intervention to which al-Awlaki was exposed can take place anywhere and against anyone if the sovereign deems it to be necessary. This is because whether such an intervention ultimately occurs does not depend upon any extant normative limitations on the authority of the sovereign—indeed, we have seen that such limitations are incompatible with the logic of preemptive security itself. Rather, it depends ultimately upon the sovereign's conjectural imaginings of potential futures that may or may not come to pass. The point, then, is that under the preemptive security regime being prosecuted through the American drone warfare program, if President Obama had decided *not* to kill al-Awlaki, this would not have been because of any juridical limitations upon his doing so; it would have instead ultimately been because the president himself simply *decided* not to—a decision which, one way or the other, is rendered absolute (in the Schmittian sense) by the logic of preemption. Thus, regardless of whether one believes that al-Awlaki's killing was legitimate, the case illustrates the sort of precarious subjectivities that are enacted by a preemptive security rationality. In particular, it shows how the law is effectively diminished as a meditative barrier between sovereign and subject in such a way that the latter is rendered perpetually vulnerable to being unilaterally and incontestably constructed as bare life if the sovereign deems it fit, regardless of who they might be or where they might be located.

This latter point—that preemption diminishes the limitations on who can be targeted for anticipatory action and/or where this targeting can occur—is an especially important aspect of the precarious subjectivities enacted by anticipatory security regimes, and can be further clarified by considering two particular details of the al-Awlaki case. The first is that al-Awlaki was an American citizen. This is significant because the Fifth Amendment to the US Constitution protects all citizens from being "deprived of life, liberty, or property, without due process of law." Accordingly, possession of American citizenship ought to shield an individual from execution by the state on the basis of a unilateral executive decree; yet al-Awlaki's killing was authorized and carried out on precisely such terms. This provides a useful illustration of the point that there can be no *a priori* restrictions upon *who* can be targeted for anticipatory intervention under a preemptive security regime. Indeed, we have seen that this is a necessary component of the latter, since deference to such juridico-normative circumscriptions as the rights of citizenship would limit the enhanced discretionary authority of the sovereign, thus compromising its capacity to render the preemptive imperative actionable. The al-Awlaki case thus shows how the imperative to tame temporal contingency trumps existing normative mediations between sovereign and subject, such that even the protections of citizenship can be disregarded in the present in the name of attempting to govern the future.

I am focusing here on the al-Awlaki killing, but it should be noted that this subordination of extant legal protections to the imperatives of preemption can be detected in counterterror practices beyond the American drone program. For instance, Matthias Leese makes a similar point in his critique of ongoing efforts to enhance the preemptive capacity of the EU's aviation security regime by creating a passenger name record (PNR) based on algorithmic, data-driven risk profiling (2014). He argues that this will inject a "fluidity" into the resulting profiles that renders the process incompatible with long-established safeguards against discrimination enshrined in EU jurisprudence, which "only unfold their full regulatory power when applied to discrimination that arises from traditional practices of profiling that sort populations on the basis of predefined individual characteristics" (500). In other words, the establishment of this particular anticipatory security practice will drastically weaken the extant legal mediations between the petty sovereigns of the aviation security regime and the mobile subjects over whom they exercise decisional authority, and this is precisely because these mediations are effectively inapplicable to the aspects of the practice designed to augment the state's preemptive capabilities. As Leese puts it, "The role of law within this assemblage . . . diminishes owing to the non-applicability of its tools"—a point that also speaks to the argument from the preceding chapter that the logic of preemption is fundamentally incompatible with the norms of a liberal legal regime (505).

The second aspect of the al-Awlaki case that illustrates the unmediated relation between sovereign and subject in the context of preemptive security is the fact that the killing took place on Yemeni territory, and outside the legally defined jurisdiction of the American executive that authorized and carried it out. This shows that the decisionist form of sovereign authority presupposed by preemptive security operates beyond legal circumscriptions relating not only to *who* one is—such as the protections of citizenship—but also to *where* one is located—such as the jurisdictional boundaries of states. This is again because the imperative to tame a radically uncertain future requires a sovereign power largely unconstrained by juridico-normative limitations, which includes those associated with present spatio-political distinctions such as state borders. Indeed, sources of potential catastrophe defined temporally rather than spatially may not neatly correspond to extant spatial coordinates. As such, efforts to preempt them by acting upon the future must disregard these arbitrary constructs of the present, since deference thereto would compromise the absolute decisional authority that is required to render anticipatory imperatives politically actionable. In this respect, the adoption of a preemptive security rationality presupposes the liberation of sovereign power from circumscription by the spatialized distinctions of the prevailing global political order and their attendant juridical limits, since it can only be effectively deployed against a fluidly indefinite future if it is able to operate across the spatio-legal rigidities that constitute that order. The practical result is that the precarious subjectivities associated with the ever-present potential of anticipatory sovereign intervention are not confined to the spatial boundaries of the state from which the intervening force derives its authority, nor to any explicitly designated "exceptional space" over which this authority exercises absolute sovereignty—such as a traditional "camp" typified by the American military prison at Guantánamo Bay. Rather, the logic of preemptive security requires that the exceptional relations between sovereign and subject that obtain in these latter spaces—in which an individual is inscribed as "bare life" and thus confronts sovereign power "without any mediation" (Agamben 2000, 41)—be enacted anywhere it is deemed to be necessary by the relevant sovereign authority, irrespective of existing spatial demarcations and the normative circumscriptions that correspond thereto. Put more simply, under a preemptive security regime, spatially constituted mediations between sovereign and subject—such as state borders, "camp" fences, or other specific delineations between an "inside" and an "outside"—are rendered effectively irrelevant with respect to determining where exceptional interventions can take place.

At this point it is again worth noting that this abstract conceptual point is complicated in practice. While in theory a preemptive security rationality must be unencumbered by extant spatio-political boundaries if it is to adequately act upon the future, the contemporary practice of preemptive security is still largely shaped by the contours of the geopolitical present. The targets of American drone strikes, for example, remain confined to a particular territorial area—the predominantly Muslim states of the greater Middle East. This suggests that the drone program operates within—and in some ways productively utilizes—extant geopolitical demarcations and coordinates in the course of implementing the United States' broader preemptive security strategy. However, this does not negate the conceptual argument being made here, which would view this empirical fact as the result of a deliberate strategic choice rather than a function of some normative circumscription on the American executive's capacity to exercise sovereign power in a preemptive manner. The al-Awlaki case supports such an interpretation, since it exemplifies precisely the sort of spatially unconstrained sovereign power that I have argued is associated with the logic of preemption, as it shows that an individual unilaterally deemed dangerous by the American executive remains equally vulnerable to violent anticipatory intervention whether he or she is located beyond or within the legal sovereign jurisdiction of the United States. Put differently, the hellfire missiles that killed al-Awlaki were fired from a drone aircraft remotely piloted by American personnel, and struck and killed an American citizen on sovereign Yemeni territory. The very act of firing these missiles as a preemptive intervention against a potential future in which al-Awlaki had engineered the proverbial next attack served to enact the subjectivity of the "camp" not within the boundaries of the American state or within a specifically delineated "exceptional space" subject to American sovereignty, but along a rural Yemeni road. This illustrates how the operationalization of preemptive security creates a condition in which any*one* can be targeted for anticipatory intervention—since the targeting process relies upon a unilateral sovereign decision based on a speculatively imagined future in which all are potential suspects—and in which those who are targeted can be subjected to a sovereign intervention any*where* they might be located. Of course, practically speaking, this latter point also implies that the intervening force possesses the material capability to exercise such broad sovereign reach. Yet the al-Awlaki case shows that the advent of the drone has given the United States precisely this capacity, and it has been reported that the United States is currently in the process of augmenting its drone arsenal with a view to developing supersonic drones that are "capable of striking targets anywhere on the globe within one hour" (Canadian International Council 2012).

By demonstrating how the preemptive exercise of sovereign power transcends limiting norms relating to a targeted individual's citizenship affiliation and spatial location, the al-Awlaki case offers a useful illustration of the connection between preemption and exceptionalism, and an archetypical example of the type of unilaterally decided, corporeally violent anticipatory action that is correspondingly made possible. In fact, the due process-free killing of a citizen by his own government represents exactly the type of exceptional intervention that is enabled—and perhaps even demanded—by the logic of preemptive security, highlighting the precarious character of political subjectivity that accompanies the adoption of rationalities based thereupon. Indeed, so long as a preemptive security rationality prevails, all political subjects within the reach of a sovereign authority operating in accordance with it are rendered perpetually vulnerable to such interventions as that which killed al-Awlaki, *even if they are never actually targeted in practice*. In other words, the very fact that the al-Awlaki killing *could* take place at all shows that this state of vulnerability to the "lightning strike" of anticipatory sovereign violence is a defining feature of the subjective experience under a preemptive security regime. To be sure, such interventions are infrequent in practice, but this does not diminish the conceptual point that such interventions are necessarily *made possible* by the adoption of a preemptive security rationality, and that this *possibility* enacts highly precarious conditions of political subjectivity. These precarious subjectivities are the result of a *de facto* elimination of juridico-normative mediation between sovereign power and political subject, which permits the former to confront the latter in much the same way as in such overtly exceptional spaces as the Agambenian "camp" or the Schmittian decisionist state. The creation of such conditions is precisely what is at stake with the adoption of preemptive approaches to (in)security governance.

CONCEPTUAL INCOHERENCES

The core line of argument developed across both this and the previous chapters of part II can be distilled down to the following: due ultimately to the political temporality embedded in the logic of preemption itself, the adoption of a security rationality based thereupon entails the creation of what amounts to an exceptional(ist) political condition, with the result that the lived experience of political subjects can be best described as precarious. The contradictory tensions inhering in the idea of preemptive security should already be somewhat apparent, since such a condition of precarious subjectivity is antithetical to the idea of "security" in its most elementary sense. To be sure, all security rationalities embody certain tensions, such that the pursuit of security in accordance with their tenets also produces *in*security to a greater or lesser extent—indeed, shedding light upon such tensions is the primary purpose of overtly "critical" approaches to the study of security. However, with respect to the notion of preemptive security in particular, these tensions are especially glaring, to the extent that the conceptual coherence of preemption as a security rationality can be brought into serious question. Indeed, the arguments developed above suggest that, by virtue of its own constitutive logics, preemptive security rationalities bring into being precisely the sort of *in*security that they are normatively premised upon diminishing. In other words, the notion of preemptive security does not merely fail with regard to its underlying normative promises to deliver a particular form of "security"; it also actively inscribes a political condition characterized by the opposite of these promises.

To unpack these points further, it is useful to take a step back and more clearly enumerate the tensions inhering in the underlying logics of preemptive security rationalities. In this respect, the discussions thus far have shown how imperatives to govern the future demand that an effectively arbitrary, life-and-death decisional authority be granted to relevant sovereign authorities. This, in turn, creates an environment in which individual citizens are perpetually vulnerable to being deemed dangerous and thus subjected to a potentially violent anticipatory intervention. While pernicious excesses where wholly innocent individuals are targeted in this way are rare in practice, the key point is that such instances *are* nevertheless *made possible* by the logic of preemptive security. Indeed, looking beyond the Mohamed Hersi and Anwar al-Awlaki cases (where questions of "innocence" are less clear), one need only consider other high-profile cases—such as the shooting death of Jean Charles de Menezes discussed in chapter 4, or the 2002 extraordinary rendition of Syrian-Canadian Maher Arar—to appreciate this point. Yet the most important issue for present purposes is the conceptual point that the institutionalization of this immanent possibility of arbitrary violence is an originary function of the logic of preemption itself, since it is ultimately traceable to the operational requirements of a political imperative to act on the future. This is problematic for the coherence of preemption as a security rationality in particular because it suggests that anticipatory security strategies imply the enactment of a condition where anyone can be targeted and subjected to violence at any time, regardless of the political circumstances. Indeed, precisely because it is an originary function of the logic of preemption itself, such an exceptional condition would emerge under a preemptive security regime even in the case of states ostensibly committed to the rule of law and the upholding of human rights norms, thus rendering even their own citizens perpetually vulnerable to these sorts of exceptional interventions—whether at the airport, at the border, in a London Tube station, on a rural Yemeni road, or elsewhere (see Amoore 2008, 115). Because such a condition of perpetual vulnerability is incompatible with the most basic understanding of security as "a condition of being protected [and] free from danger," this seriously challenges any claim that preemptive strategies are able to deliver anything that might be understood as "security" (Der Derian 2009, 152).

These conceptual problems are exacerbated by the explicitly precautionary ethos which underpins contemporary manifestations of preemptive security and the attendant bias toward acting *now*. As we have seen in chapter 4, this leads to an emphasis upon immediate action over moderated restraint, thus ensuring a high level of anticipatory activity and a concomitantly increased likelihood that inevitable errors will result in excessive and unjustified violence. In other words, anticipatory political strategies such as preemptive security prioritize action over inaction (de Goede, Simon, and Hoijtnik 2014, 419), and as a consequence "do not take seriously enough the downside of getting things wrong" (Runciman 2004). Thus, in addition to increasing the likelihood that individuals will be wrongly targeted, this has the effect of ensuring that incidents such as the Menezes shooting—in which an innocent man was killed by petty sovereigns acting under a preemptive rationality—can be framed as mere "accidents" or "mistakes" that, while regrettable, are an inevitable price to be paid for the putative protections of a preemptive security regime (Vaughan-Williams 2007, 183). By placing the blame for such excesses on the unique circumstances of each case, this obscures the crucial point that—in a manner reminiscent of Virilio's (2007) notion of the "integral accident"—they can be traced back to the logic of preemption itself, which is what ultimately makes them possible. It therefore becomes quite difficult to resist the ongoing embrace of anticipatory security rationalities by political elites, since particular irruptions of the exceptional violence that such rationalities necessarily make possible can be discursively papered over in these sorts of ways. The potential implications for the experience of political subjectivity, and thus everyday life, under a preemptive security regime are therefore quite significant—particularly in the context of liberal democratic polities.

However, these considerations do not merely imply that preemptive security rationalities are incapable of delivering a condition that can be understood as "security"; perhaps more important, they also suggest that anticipatory strategies in fact reproduce precisely the form of *in*security that they are ultimately premised upon diminishing. In this regard, consider exactly what it is that preemptive security rationalities seek to achieve. Put most simply, they promise, above all, to protect human subjects against *potentially forthcoming harms*, since the threats against which they are framed are located in a future that may or may not come to pass. Put another way, preemptive security rationalities are premised upon the reduction—and, ideally, the elimination—of a certain vulnerability caused by exposure to *potential violence* represented by the possible future irruption of catastrophe. The problem is that by responding to an uncertain future by enacting a politics of exceptionalism in which sovereign authorities can target subjects for anticipatory intervention on the basis of speculation and conjecture, preemptive security replaces the initial form of vulnerability that it seeks to diminish—to the irruptive threat posed by a radically uncertain future—with another form of vulnerability—to the will of a decisionist sovereign authority that can unilaterally intervene suddenly and violently in an effort to govern that future. Indeed, the *potential violence* represented by vulnerability to the "next terrorist attack" is subjectively little different from the *potential violence* represented by vulnerability to indefinite detention, assassination by drone strike, summary execution in a London Tube station, or any other such exceptional exercises of sovereign power that are made possible by a politics of preemptive security. Thus, while the latter is normatively premised upon protecting against the potential violence of an uncertain future, the mechanisms by which it does so create a lived present fraught with an alternative form of potential violence in the form of perpetual vulnerability to an extrajudicially decided lightning strike by state security apparatuses. Indeed, this possibility is part and parcel of preemptive security, which accordingly places citizen-subjects in arguably as precarious a position as their untempered exposure to the potential catastrophes lurking in the future's unknowable depths.

The adoption of preemptive security rationalities bring into being precisely the sort of condition that they are premised upon eliminating, and so, like many governmental innovations of the post-9/11 global security climate, it "ends up producing, reproducing, and regenerating the very thing it seeks to disarm" (Derrida 2003, 99). This calls into question the coherence of the idea of preemptive security itself, and poses a serious challenge to its viability as a policy strategy for contemporary (in)security governance. This conceptual incoherence, or fundamental impossibility, of preemptive security is perhaps the central conclusion of part II of this book, and the next section suggests that reading this idea through Derrida's concept of "autoimmunity" is a particularly effective way to emphasize it.

"AUTOIMMUNITY"

The quotation from Derrida in the preceding paragraph was deliberately chosen, as it is taken from his own critical engagement with the Western response to the attacks of 9/11—a response that was spearheaded by the types of anticipatory strategies with which this book is concerned (Derrida 2003). In context, the quotation is part of his broader description of how the sociopolitical dynamics unleashed by this response represent a concrete illustration of his notion of "autoimmunity" (Derrida 2005, 2003). This concept is central to Derrida's later "political" writings, and although it has been widely discussed across a number of disciplines, its appropriation by the international relations and (critical) security studies literature has remained rather limited.[2] In this section, I consider how the idea of autoimmunity offers a useful conceptual lens through which to theorize the conceptually problematic character of preemptive security, since it concisely describes the sorts of internally contradictory tensions that the previous section has identified, and thus emphasizes how their problematic practical effects are in fact an originary function of the logic of preemption itself. This point is crucial, since it can productively serve as the conceptual foundation for a serious policy critique of the emerging consensus around the utility of preemptive security strategies in particular and anticipatory governance rationalities more generally.

Derrida's Idea

We must begin by elaborating upon what precisely Derrida meant when speaking of "autoimmunity." The term itself is, of course, derived from the vernacular of medical science,[3] where it describes a category of disease in which the body's own immune system "misrecogniz[es] parts of itself as other than itself and then seeks to eliminate these unrecognized and hence antagonistic aspects of itself." An autoimmune response constitutes "a living contradiction," as the very mechanism whose purpose is to protect the body—the immune system—ends up turning against the body and ultimately harming it (Cohen 2004, 8). Derrida appropriates this idea and extends its application to the realm of the social, asserting that in fact "all kinds of beings, from discourses to institutions" are subject to what he terms "a general logic of autoimmunity" (Derrida 2002; Naas 2006, 25).

In elaborating on this "general logic," however, Derrida modifies the meaning of autoimmunity from its original medical formulation in subtle, yet important, ways. In particular, he places particular emphasis on its relation to the root concept of "immunity" by specifying that *auto*immunity only emerges subsequent to an original "immunizing gesture" (Arfi 2010a, 246; 2010b, 312; Naas 2006, 34). For Derrida, any attempt by a social entity—such as a political community—to "immunize" itself against a putatively threatening "other"—for example, through the construction and defense of a sovereign boundary—necessarily renders that entity subject to the "illogical logic" of autoimmunity (Derrida 2005, 35–36). Autoimmunity is thus originarily present in any immunizing move, as "there is no immunity without autoimmunity"—the one is part and parcel of the other (Derrida 2003, 159). Moreover, because the autoimmune process is precipitated by the immunizing gesture itself, Derrida asserts that it is more specifically the *immunity* of the entity that is destroyed by the logic of autoimmunity. In other words, the process of autoimmunity represents the destruction not of the entity itself—although this may ultimately follow, and often does—but more specifically of the entity's own immunitary protections. This particular specification is important for Derrida, as it is made clear in two of the more eloquent descriptions of his understanding of autoimmunity. In *Rogues*, for instance, he describes it as "this strange illogical logic by which a living being can spontaneously destroy, in an autonomous fashion, the very thing within it that is supposed to protect it against the other, to immunize itself against the aggressive intrusion of the other" (2005, 123). Similarly, in *Philosophy in a Time* *of Terror*, he describes autoimmunity as "that strange behaviour where a living being, in quasi-*suicidal* fashion, 'itself' works to destroy its 'own' immunity" (2003, 94, emphasis original; see also Borradori 2003, 20).

In accordance with this formulation, Derrida emphasizes that in the socio-political context, the operation of this logic can lead an original immunizing gesture to ultimately bring into being precisely that which it is premised upon protecting, or "immunizing," against (Derrida 2003, 99). In other words, since it is the immunitary protection itself that is compromised, the types of condition(s) that this protection seeks to preclude are ultimately made manifest by the autoimmune process. Derrida claims that any attempt to foreclose against what he calls "the coming of the other"—by which he means any "event" situated beyond the realm of knowability/calculability—inevitably leads to an "exposure" to precisely that "other" through the logic of autoimmunity that is always already immanent in such a move (2005, 152–53; Arfi 2010a, 246; Gasché 2004, 295).[4] In more concrete terms, these ideas have been deployed by Derrida himself, along with numerous others, with respect to the actions of the liberal West in the wake of the 9/11 attacks (Cheah and Guerlac 2009, 12–13; Derrida 2005, 40; 2003: 94–99; Miler 2009, 128; Mitchell 2005, 919; Naas 2006, 31). The arguments in these works vary in nuance and sophistication, but they generally focus upon the enactment by such polities of various illiberal practices and emphasize how the very politico-juridical elements that ostensibly "immunize" liberal democratic states and their citizens against such ills as arbitrary sovereign violence and human rights violations are therefore transgressed in the name of securing those very elements through an ongoing "War on Terror." Such arguments thus highlight the key aspects of the concept of autoimmunity by emphasizing both that it is "immunity" itself which is compromised by the autoimmune process and that the practices that trigger the process ultimately end up bringing into being the very type of condition they aspire to secure (or "immunize") against (Derrida 2003, 94).

Preemptive Security as Autoimmune

In light of both Derrida's formulation and the arguments developed in this and the preceding chapters of part II, the idea of preemptive security can be usefully read in terms of "autoimmunity." Indeed, not only do preemptive security rationalities constitute precisely the sorts of social processes that Derrida claims are subject to the logic of autoimmunity, but the conceptual tensions identified in the previous section are also an excellent example of that logic being borne out. In this respect, recall from earlier chapters that the advent of anticipatory rationalities of government was precipitated by an apparent need for reformulations of sovereignty aimed at more actively taming the radical uncertainty of the late-modern world. As a conspicuous manifestation of this imperative, the adoption of preemptive security rationalities can be understood as exactly the sort of "immunizing gesture" of which Derrida speaks, since they are normatively premised upon diminishing vulnerability to potential violence by "immunizing" against the catastrophic potentialities deemed to inhabit the unknowable depths of the future. Moreover, such potentialities—typified by the specter of the "next terrorist attack"—constitute the very type of incalculable "event" or "other" against whose coming a Derridean "immunizing gesture" seeks to foreclose (Derrida 2005, 135, 144, 148, 152; Gasché 2004, 295). The idea of preemptive security as a sociopolitical project is thus exactly the sort of process to which the "illogical logic" of autoimmunity is claimed by Derrida to apply.

Derrida also emphasizes that this logic is activated by the original immunizing gesture itself, and this description also applies to the conceptual tensions inherent to the idea of preemptive security. Indeed, it is the attempt to (re)assert control over temporal contingency—which constitutes a prototypical "immunizing gesture"—that creates the imperative to act upon the future, which in turn prioritizes the imagination and thus enacts an exceptionalist politics that ultimately compromises the preemptive security project by bringing into being a condition of precarious subjectivity akin to that which it aims to diminish. In other words, just as in Derrida's logic of autoimmunity—in which the autoimmune process is triggered by the original immunizing gesture—the constitutive contradictions of preemptive security are activated at precisely the moment that it is decided to make the immunizing gesture of turning the idea of preemption into a practical rationality. From this perspective, part II of this study can be read as an extended description of how the autoimmunitary logic of preemptive security unfolds.

Finally, Derrida argues that the logic of autoimmunity ultimately leads to the creation of that which the original immunizing gesture was aimed at protecting against. This is precisely what results from the adoption of preemptive approaches to (in)security governance, since the political exceptionalism required by the underlying imperative to act upon the future merely replaces one form of precarious subjectivity with another. Preemptive security rationalities are normatively premised upon diminishing vulnerability to the potential violence of future catastrophe; however, this aim is pursued by ultimately institutionalizing vulnerability to the potential violence of a radically decisionist sovereign authority impelled to intervene preemptively on the basis of conjecture and speculation. Thus, in a manner congruent with the Derridean autoimmune process, preemptive security's internal conceptual tensions not only preclude it from ultimately providing the "immunity" it promises but in fact lead it to bring into being precisely the sort of condition that it is premised upon eliminating.

Derrida's concept of autoimmunity therefore provides a very useful lens through which to understand the conceptual tensions inherent to the idea of preemptive security. Not only does it succinctly capture how the latter fails to live up to its normative promises, and in fact ends up creating that which it aspires to eliminate, but it also conveys the crucial point that this aporiatic contradiction is originarily embedded in the logic of preemption itself. This is important because in addition to illustrating how preemptive security rationalities operate through manifestly illiberal practices that ultimately exacerbate the *in*security of political subjects—an argument that has been made at length in the existing critical security studies literature—part II of this book has also sought to emphasize that this is not merely due to poor implementation by specific political authorities in particular states, but rather is a function of the very idea of anticipatory governance itself. Describing preemptive security rationalities as "autoimmune" conveys both of these points clearly and succinctly, as it invokes a familiar medical condition where the cohesion and integrity of an entity is compromised by a potentially fatal internal contradiction, and in which the mechanism that is supposed to protect the entity from exogenous harm in fact creates such harm endogenously.

Moreover, with respect to this book's broader contributions, an emphasis upon both of these points together is important. This is because although this study is primarily conceptual in nature, these points together can serve as the conceptual foundation for a more policy-oriented critique of preemptive security in particular, and anticipatory governance more generally. In this respect, they suggest that the problematic practices associated with preemptive security rationalities cannot be remedied by simply modifying existing policies or implementation strategies. Rather, because these problems are fundamentally inherent to the logic of preemption itself, they will continue to emerge whenever some form of preemption is the overarching aim of a particular policy regime, regardless of the way in which it is practically implemented. This undercuts important lines of defense against critics of preemptive security examples, since it renders untenable the claim discussed in the previous section that such violent excesses as the Menezes shooting are mere aberrational "mistakes" unique to the circumstances of a particular case. It instead makes clear that the continuous *possibility* of such excesses is a necessary condition of preemptive approaches. Viewed from this perspective, an argument for further entrenching anticipatory security strategies becomes much more difficult to support, as an equally compelling argument can be advanced that the wholesale embrace of preemptive security will merely replace one form of *in*security with another. Understanding preemptive security in terms of "autoimmunity" may therefore help policymakers currently enamored with the idea to more thoroughly scrutinize its viability.

Thus, in addition to being analytically useful in terms of concisely capturing the key conceptual points developed in part II of this book, framing preemptive security in terms of Derrida's notion of "autoimmunity" creates discursive space for further critical interventions that explicitly question its practical validity and policy legitimacy—particularly in the context of the liberal democratic states whose core juridico-political tenets are explicitly challenged by the exigencies of preemption, yet whose policymaking elite remain strong supporters of it. Indeed, the al-Awlaki killing—which was undertaken by the most powerful of such states—offers an archetypical example of the autoimmune process in action, since the political process through which the imagined threat he posed was preempted required enacting a form of sovereign authority whose existence poses an equally serious threat of potential violence to those it is ostensibly trying to protect.[5] Understanding preemptive security in terms of autoimmunity clearly emphasizes such tensions, thus lending greater urgency and legitimacy to attempts to resist the emerging consensus in favor of anticipatory governance strategies that characterizes contemporary liberal democratic polities.

CONCLUSION

Under the broader aim of considering what is at stake with the ongoing proliferation of anticipatory governance strategies in the current global political moment, a key aim of this book is developing a detailed critical interrogation of the notion of preemptive security—a prototypical manifestation of this trend. Toward this end, chapters 3, 4, and 5 have used the politics of preemptive security to respectively consider what preemption can be understood to mean in the contemporary context, how it operates as a political rationality, and the paradigm of political authority it requires to be institutionalized. The task of this sixth chapter has been to take stock of the arguments developed in these discussions and critically consider what they imply with respect to more practical implications of adopting preemptive approaches to (in)security governance. In this respect, it was argued that the reliance of a preemptive politics upon what amounts to an exceptionalist form of sovereign authority leads security strategies premised thereupon to enact a political condition characterized by a precarious subjective experience. In particular, the types of practices that are necessarily made possible by a preemptive security rationality place those subjected thereto in a position of perpetual vulnerability to what amounts to arbitrary, extrajudicial interventions by the agents of sovereign power. Having illustrated this point through an in-depth discussion of the targeted killing of Anwar al-Awlaki, it was subsequently argued that the necessary creation of this status of vulnerability to such interventions—regardless of the (in)frequency of their occurrence—seriously undercuts the coherence of preemptive security as an idea. This is because the latter is premised upon countering precisely that sort of precarious subjectivity by insulating against the catastrophic contingencies of a radically uncertain future. One form of vulnerability to sudden irruptions of violence is thus replaced with another.

The notion of preemptive security can be productively read through the lens of Derrida's concept of autoimmunity, since doing so emphasizes the degree to which these contradictory tensions are intrinsic to the logic of preemption itself. This suggests that the problematic practices associated with preemptive security can be most effectively mitigated by resisting the broader trend toward anticipatory governance, rather than through any modification to the way such strategies are implemented in particular contexts. Understanding preemptive security in terms of autoimmunity can prove a useful conceptual departure point for developing a more policy-oriented critique of anticipatory forms of governance more generally. Indeed, the widespread proliferation of preemptive security strategies represents merely one manifestation of a broader global move toward anticipatory governance based on a spectrum of future-oriented logics from risk to precaution and beyond. The critical insights developed in part II of this book can thus be extrapolated from the security realm to address the more general governance trends that have characterized wider societal responses to the apparent "end of certainty." This theme will be taken up in the conclusion.

NOTES

1. Taking this line of argument a step further, it might be concluded that the advent of the American drone warfare program—with its combination of rapid global reach and exceptionalist politics of operation—has the effect of turning entire countries such as Yemen, or even the globe itself, into what amounts to an Agambenian "camp." While perhaps somewhat excessive, such an argument nevertheless highlights the very real extent to which a preemptive security rationality—of which the drone program is one high-profile and sophisticated manifestation—renders the experience of political subjectivity quite precarious on a very broad scale.

2. Notable exceptions in this regard are Arfi 2010a, 2010b, and Vaughan-Williams 2007.

3. Although, as an interesting aside, the medical sciences appropriated the root idea of "immunity" from the social realm, where it originally referred to those "who were in one way or another exempt from the ordinary citizen's obligations"—a meaning that is residually present today in the notion of "diplomatic immunity." Indeed, the entire medical metaphor of the "immune system"—so common today—is in fact derived from the social world, based as it is on the idea of foreign invaders that must be actively confronted and eliminated by an internally constituted force (see Miller 2009, 124; Mitchell 2005, 917).

4. Derrida's detailed theoretical argument as to exactly why this is so is developed at length throughout *Rogues* (2005); however, the characteristic complexity of his claims renders even a cursory summary far beyond the scope of this chapter. For usefully concise discussions of this question, see Cheah and Guerlac 2009, 13–17, and Naas 2006, 20–26.

5. This point suggests that the fact that Derrida himself develops the theory of autoimmunity in specific reference to the broader concept of democracy is no coincidence, as the auotimmunitary turn toward anticipatory governance has been a prominent feature of contemporary liberal democratic politics (Derrida 2005).

Chapter Seven

Conclusion

Anticipatory Governance, Liberal Democracy, and the Politics of the Future

Recall from the introduction that this book was concerned with undertaking three overarching tasks: first, to emphasize that foregrounding temporality more generally and futurity more specifically can elicit productively critical insights into contemporary political and social issues, and to provide a book-length illustration of how this might be done; second, to establish the basis for a conceptual critique of the sorts of anticipatory governance rationalities whose global proliferation is a primary reason for focusing on temporality and futurity in this context; and third, to do so by developing a comprehensive critical theorization of one particular example of such forms of governance—namely, the logic of preemptive security that has infused the prosecution of the global War on Terror. The pursuit of these tasks has not directly paralleled the preceding chapter divisions, and each has been given varying emphasis as the book has progressed; however, by this point it is hoped that the connections between these lines of inquiry have become clear and, perhaps more important, been adequately explored. Let us therefore briefly revisit how they have been addressed in the preceding pages.

With regard to the first task, the key point is that all of the core arguments of this study—whether relating to the conceptualization of sovereignty or the politics of drone warfare—were framed through questions of temporality and futurity. The book's arguments can be collectively understood as an illustration of the type of political analysis that results from adopting a "temporal lens," since they were ultimately made possible by placing time "front and center" (Ancona et al., 645). Moreover, by discussing how shifts in the broader political imagination and concomitant developments in the realm of political praxis have (re)inscribed temporal contingency as a significant political problem, the arguments also helped reiterate why this sort of "bringing time in" to the study of (global) politics is increasingly important methodologically. Indeed, a failure to do so would preclude an adequate understanding of the increasingly "temporalized" realm of societal governance.

With respect to the second task, the book established the basis for an in-depth conceptual critique of the anticipatory governance strategies that have resulted from the aforementioned shifts in political imagination and praxis by illustrating how such strategies require forms of political authority that conflict with the norms and principles of liberal democracy—a trait that stems directly from their emphasis upon actively governing the future. While these points were developed through an in-depth exploration of one particular area—the preemptive security strategies of the War on Terror—the abstractly theoretical character of the analysis suggests that logics of anticipatory governance more broadly imply significant modifications to the way political power and authority is organized and exercised. As will be discussed in the next section, this point has broader implications for the democratic legitimacy of such political rationalities.

The completion of the third task served as the mechanism through which the second was pursued, as part II of the book in particular developed an in-depth account of the preemptive politics of (in)security which illustrates how temporally inflected approaches to the problem of transnational terrorism entail a shift toward a politics of exceptionalism—a shift that is once again traceable to a constitutive emphasis on taming an uncertain future. By attempting to preempt future catastrophes through anticipatory action in the present, preemptive modes of (in)security governance actively manipulate our relation to time by "making the future present." This, in turn, has the effect of compressing the timescape of political decision-making and requiring "affective facts" or "gut feelings" rather than verifiable evidence to serve as the primary basis for such decisions. This reliance upon speculation about potential futures that are to be acted against confers upon the relevant sovereign authority—which is most often the executive branch but can also include mundane bureaucratized authorities, or "petty sovereigns"—a radically enhanced degree of decisional discretion. The democratic character of the polity in question is effectively compromised, as a paradigm of political authority conspicuously reminiscent of that which obtains in a so-called state of exception is created wherein the juridico-normative mediations between sovereign and subject are diminished and a condition of precarious subjectivity is enacted.

This latter line of argument comprises a primary contribution of the book, since it provides a comprehensive conceptual picture of a particularly conspicuous manifestation of anticipatory governance. This may subsequently serve as a theoretical foundation upon which future critical work relating to more specific examples of preemptive security can be based. Moreover, by proposing and unpacking an originary conceptual link between the logic of preemption and the politics of exceptionalism, this account also contributes to ongoing discussions relating to the proliferation of exceptionalism in post-9/11 global politics—a central theme of contemporary critical security scholarship. Indeed, it suggests that this trend can be explained at least in part by the broader turn toward anticipatory governance that has characterized our ongoing response to the contingencies of late modernity. While they are important elements of this book's contribution to ongoing scholarly debates, these narrower disciplinary contributions hint at the relevance of the book's core insights relating to preemptive security for the notion of anticipatory governance more generally, to which I now turn.

ANTICIPATORY GOVERNANCE AND LIBERAL DEMOCRACY

While the myriad security practices based on the concept of preemption (or some variation thereof) that have been adopted across the globe in recent years differ widely in their specific characteristics, all can be understood to reflect an emerging consensus that the most pressing security problems—chief among which is transnational terrorism—can be best addressed through future-oriented modes of governance ultimately premised upon (re)asserting a degree of control over the unfolding of the future. In this sense, the rise of preemptive security is not wholly a unique manifestation of idiosyncratic post-9/11 (in)security politics, but instead can be situated within the wider "shift from responding to past events to preventing future harms" that constitutes "one of the most significant but unnoticed trends in the world today" (Dershowitz 2006, 7). As such, the preceding chapters' arguments about preemptive security also carry lessons for the wide variety of anticipatory governance rationalities being deployed in areas other than the ongoing efforts to confront transnational terrorism. The most important consideration here is, again, that the operationalization of anticipatory governance strategies requires the epistemic prioritization of imagination, which, in turn, entails significant modifications to the organization and exercise of political power that enact what amounts to an exceptionalist politics. Indeed, it is in this respect that the arguments developed in part II have relevance beyond the security realm, since it is precisely this sort of ultimately counter-democratic reformulation of sovereign power that is at stake with the broader proliferation of anticipatory societal governance, of which the rise of preemptive security is but one manifestation.

Of course, this is not to say that the particular form of exceptionalist politics that this book has identified as part and parcel of a preemptive approach to governing terrorism—in which the prerogative to make juridically unbound, life-and-death decisions on the basis of speculative knowledge is vested in relevant sovereign authorities—will necessarily emerge wherever and whenever a form of anticipatory political rationality is implemented. The actual practices that result from adopting such governance strategies will certainly vary considerably with respect to the particular issue areas and the sorts of problems to which they are considered a proper policy response, and may even be incorporated into existing politico-juridical paradigms in a manner that obscures the exceptionalism inherent in the logic of anticipatory governance. For example, in examining the emergence of so-called vital systems security in the United States during the Cold War period and its subsequent proliferation thereafter, Stephen Collier and Andrew Lakoff stress that this particular paradigm of anticipatory governance—which has been extended beyond its original domain of national security to include a wide variety of potential emergencies across many contemporary liberal democratic states—"does not require recourse to extraordinary executive powers" (2015, 19). Instead, like the notion of counter law discussed in chapter 5, the discourse of vital systems security effectively folds the exceptionalities inherent to anticipatory governance into the extant liberal order. Indeed, Collier and Lakoff make a convincing case that, through the wide array of preparedness measures, scenario planning exercises, and emergency management protocols that have become core components of contemporary liberal governance, this anticipatory paradigm "provides a ready-to-hand toolkit for administering emergencies as a normal part of constitutional government" (25). In this sense, therefore, anticipatory governance need not always be premised upon the exceptional exercise of sovereign power (Adey, Anderson, and Graham 2015). However, conceding this point does not necessarily negate the primary *conceptual* argument of part II, which holds that the *logic* of anticipatory governance requires what amounts to a politics of exceptionalism to be practically operationalized due to its emphasis upon governing an ultimately unknowable future. Indeed, as we have seen in the discussion of counter law in chapter 5, the enfolding of exceptional practices into an existing normative order does not so much evacuate these practices of their exceptionalism as it evacuates the polity of its liberal democratic character.

It is also the case that anticipatory practices implemented in other issue areas may not be as normatively fraught as those which have resulted from the rise of preemptive security in the post-9/11 era, and they may legitimately constitute the most appropriate response to the particular governance problem in question. In the realm of environmental governance, for example, the specter of catastrophic climate change may be interpreted as demanding a precautionary approach premised upon taking concrete action to avoid imagined worst-case scenarios that could be quite improbable. However, despite the inflated rhetoric of right-wing political discourse that might claim otherwise, the resulting practices are likely to take the form of more stringent government regulation of emissions and pollutants rather than the assumption and exercise of the prerogative to target and kill citizens. But again, the broader conceptual point remains that, irrespective of the particular context, governance *decisions* under an anticipatory rationality must be based upon imagined future potentialities, and this has the effect of enhancing the discretionary subjectivity of the deciding authority and undermining the coherence of liberal political and legal norms in that context. Any political move toward the adoption of anticipatory governance rationalities must recognize that this sort of modification to the architecture of political power will follow in at least some form, and must therefore reckon with the potential implications.

This is a particularly important point to consider in the context of self-identified liberal democratic states. Indeed, on the one hand, the arguments developed in this book have extensively demonstrated how the underlying logic of anticipatory governance operates in tension with the constitutive juridico-political principles associated with a liberal democratic society. Yet, on the other hand, it is precisely the polities that identify as such that are in the vanguard of formulating and implementing explicitly anticipatory approaches to pressing political problems—both in the security realm and beyond. Consider the European Union (EU) in this respect. The core political requirement of the Copenhagen Criteria for membership in the Union is that a state possess "stable institutions guaranteeing democracy, the rule of law, human rights, and respect for and protection of minorities" (European Union 2013). The preceding chapters have suggested that all four of these may be undermined to at least some extent by the adoption of explicitly anticipatory forms of governance. Yet the European Union has also publicly committed to a policymaking process which explicitly includes one particular well-known anticipatory approach to policymaking—the "precautionary principle," which was described at length in chapter 2 and, while differing in important ways, has exerted significant influence on the preemptive security thinking of the post-9/11 era discussed in the subsequent chapters (European

Commission 2000). According to Sunstein and Hahn, "The European Union has taken a leadership role in promoting the precautionary principle as a basis for making decisions" across a range of issue areas, including environmental governance, food policy, consumer protection, and technological development (2005, 1). The conceptual arguments developed in this book suggest that this embrace of a precautionary approach to policymaking entails significant political implications with which the European Union must reckon if it is to maintain its commitment to core liberal democratic political and legal norms. For example, we have seen how anticipatory governance rationalities grant significant subjective discretion to decisional authorities who are often unelected "petty sovereigns"—a point that parallels some of the most trenchant critiques of the EU project, which are rooted in claims about its excessive and undemocratic bureaucratization. Again, the point is not that anticipatory governance strategies are never legitimate in a democratic polity, but rather that the ways in which such approaches alter the organization and exercise of political power must be taken seriously, since these modifications can undermine the constitutive principles and norms of liberal democracy in crucial ways.

Moreover, while this book has focused on the originary link between the logic of anticipatory governance and the politics of exceptionalism, others have highlighted additional anti-democratic implications that have recently appeared. Louise Amoore (2014b), for instance, draws attention to the seemingly absurd spectacle of the Italian government charging, convicting, and imprisoning a group of seismologists for inadequately calculating the risk of the earthquake that struck the town of L'Aquila in April 2009, killing over 300.[1] Amoore attributes this episode to "an emerging form of responsibility to calculate the incalculable" where "failing to infer, intuit, imagine, and extrapolate from the available data" renders those delegated with such responsibilities culpable for any deleterious consequences that may follow (424). In other words, anticipatory imperatives have become such powerful drivers of contemporary societal governance that the "experts" tasked with their implementation—a responsibility that, as we have seen, can turn such figures into Butlerian petty sovereigns endowed with significant power—are themselves subject to exceptional interventions if they are deemed by the state to have failed to adequately facilitate preemption. The jailing of scientists solely for a perceived failure to accurately assess the risk of a possible future event is not the sort of practice generally associated with a democratic polity, and while the L'Aquila case remains something of an extreme aberration, Amoore's analysis—along with the core arguments of this book—indicates that such an outcome is logically congruent with prioritizing anticipatory modes of action in the context of a state's governance tasks.

This incident suggests that the counter-democratic implications of adopting anticipatory strategies extend beyond even the sorts of sovereign exceptionalisms and precarious subjectivities discussed above, as these implications may also include such seemingly incongruous practices as an ostensibly free society holding scientists criminally responsible for deaths caused by so prototypically unpredictable an event as an earthquake. This further signifies the potential for a robust conceptual critique of anticipatory governance from the normative position of liberal democratic politics—a project to which this book's arguments can contribute.

A broader conclusion might be that any policy decision to implement a form of anticipatory governance must be taken with great caution, with the potential consequences for democratic principles weighed carefully against the purported benefits of ostensibly reasserting some degree of control over the unfolding of the future. Such judicious reflection might, for instance, recommend a rollback of the preemptive security strategies characterized by targeted killing, while also endorsing anticipatory action on environmental matters, where intervention would not take the form of effectively arbitrary sovereign violence. In short, while anticipatory governance rationalities may represent an important political tool in ongoing efforts to confront the key problems of late modernity, this book has attempted to show that such strategies carry with them significant ethico-political baggage, and must therefore be deployed with great care if this tool is to be used effectively and responsibly.

TEMPORALITY, FUTURITY, AND THE STUDY OF (GLOBAL) POLITICS

In addition to drawing attention to these more "policy-relevant" questions, this book has also addressed itself to more purely academic questions relating to the analytical utility and critical potential of foregrounding temporality and futurity in the study of (global) politics. In this respect, the key point is that the core insights leading to the conclusions about the anti-democratic character of anticipatory governance were a direct result of such a foregrounding. Consider in this respect how chapter 3's description of the post-9/11 conceptualization of preemption as applied to the realm of (inter)national security shows that the latter can only be properly understood by foregrounding its temporal aspects. Indeed, because it inscribes radical uncertainty as the basis for, rather than an impediment to, anticipatory action, it is not simply a mechanism for avoiding particular imminent possibilities, but a rationality for controlling the unfolding of time more broadly. Similarly, chapter 4's critical account of preemption's political logic also relies upon the foregrounding of time, since this logic is best understood by considering how it actively manipulates our relation to time by effectively "making the future present." Relatedly, the originary link between preemption and exceptionalism discussed at length in chapter 5 was expounded by unpacking the political temporality of preemption and considering how its prioritization of the future inscribes the imagination as the epistemic foundation for anticipatory political action. Finally, because it enacts what amounts to an exceptionalist political paradigm, it is this temporal imperative to control the future that is the ultimate source of the precarious subjectivities that chapter 6 argued are part and parcel of preemptive security rationalities. Chapter 2's description of the ongoing "temporalization" of the political further highlights the analytical importance of foregrounding the questions of time and futurity by demonstrating how explicitly temporal problems relating to the future's radical uncertainty have been cast as pressing political issues, leading to the proliferation of anticipatory political rationalities in response (Opitz and Tellmann 2015, 111). The same chapter's temporally inflected reading of the concept of sovereignty also makes these points by showing how the question of time and its control fundamentally underpins this most elemental of political concepts.

This book's primary lines of argument collectively suggest that employing a temporal lens is key to adequately understanding phenomena that are of crucial importance to contemporary questions of societal governance, since it can enable innovative readings of these phenomena that "re-open assumptions that have grounded our political thought" (Edkins and Vaughan-Williams 2009, 2). In addition to the critique of preemptive security strategies and anticipatory governance rationalities which speaks to questions of political praxis, therefore, the arguments developed here also speak to disciplinary questions relating to the scope and framing of academic inquiry in the study of politics by demonstrating that adopting a temporal lens is both intellectually possible and analytically productive. Indeed, doing so will only enhance the critical capacity of scholars to make sense of the emerging complexities of the broader twenty-first-century social world.

"FUTURE" DIRECTIONS

It is important to note, however, that this book represents only one example of the sort of scholarly intervention that is made possible by adopting a temporal lens for the analysis of societal governance. Future work might productively begin where this study has left off, by exploring in further depth how anticipatory strategies have been adopted and deployed in other issue areas—such as environmental, financial, or public health governance—while paying particular attention to the ways in which this has affected the organization and exercise of political power in those particular contexts. Such work would add further nuance and empirical depth to this book's primarily conceptual claims about the counter-democratic character of anticipatory governance, since these claims have been largely developed through the in-depth interrogation of one prominent manifestation thereof. Such interventions would also contribute to developing a more comprehensive knowledge base regarding the potential costs and benefits of implementing anticipatory strategies across a variety of policy areas. This would help to ensure that such strategies are employed in the most responsible way possible.

To conclude, however, it is worth considering how the responsible implementation of anticipatory governance can also be cultivated through efforts to modify certain aspects of the political imagination that have fuelled the widespread embrace of such strategies as the most appropriate solutions to some of the more pressing political problems of our time. For instance, this trend might be restrained if the latent Machiavellian imperative to tame *fortuna* can be relaxed and the uncertainty of the future recast not simply as an always potentially imminent catastrophe that must be controlled, but as an unavoidable condition of the human experience that must be accepted and lived with. In other words, if we can come to terms with the ineradicability of radical temporal contingency, and thus more fully reconcile ourselves to our ultimate inability to control the unfolding of the future, the impetus to adopt political rationalities concerned with doing just that may diminish—and along with them the temptation to trade the liberal democratic principles of the present for the ostensible capacity to more effectively govern the future. Such a stance need not deny all agency to humanity in the face of time's march by returning the future to the purview of the gods; however, such a dose of temporal humility might help foster a political imagination that no longer views taming uncertainty as an imperative whose urgent pursuit can override competing imperatives relating to the organization and exercise of political power.

Interestingly, Machiavelli himself can be read as expressing a similar outlook in his celebrated discursion on *fortuna* in chapter 25 of *The Prince*. Here he famously contends that "since our free will must not be eliminated, I think it may be true that *fortuna* determines one half of our actions, but that, even so, she leaves us to control the other half" (Machiavelli 1995, 44). This passage is often read with an emphasis on the latter point about humanity's ability to actively resist the vicissitudes of time through *virtùous* action. Yet the prior point conceding that this ability is nonetheless inherently limited is also crucial, since it suggests that these limitations must be accepted, such that our efforts to tame time do not supersede all other concerns. In other words, in perhaps the most renowned work that addresses the politics of time, Machiavelli can be read as implying that while it is prudent to do our best to control what aspects of the future we reasonably can, we must also recognize that there are some contingencies that are beyond our capacity to tame through deliberate governance mechanisms. Political elites, policymakers, and public thought leaders might do well to heed this idea by judiciously limiting calls for anticipatory intervention to instances where the problem fits under the first category rather than the second. Indeed, perhaps the central lesson of this book is that if the "quest for time control" becomes so powerful a motivation in political thought and action that governing authorities overreach in their attempts to achieve it, the results may end up seriously undermining precisely those abstract principles and practical ways of life that anticipatory actions are premised upon preserving in the face of time's continuously unpredictable unfolding.

NOTE

1. Their convictions were ultimately overturned in November 2014.

Bibliography

Aalberts, Tanja, and Erna Rijsdijk. 2011. "Mobilizing Uncertainty and Responsibility in International Politics and Law: Guest Editors' Introduction." *Review of International Studies* 37: 2157–61.

Aalberts, Tanja, and Wouter Werner. 2011. "Mobilizing Uncertainty and the Making of Responsible Sovereigns." *Review of International Studies* 37: 2183–2200.

Ackerman, Spencer. 2012. "Here's Why the Government Thinks It Can Kill You." *Wired*, March 5. http://www.wired.com/dangerroom/2012/03/holder-targeted-killing/.

———. 2015. "No-Fly List Uses 'Predictive Assessments' Instead of Hard Evidence, US Admits." *Guardian*, August 10. http://www.theguardian.com/us-news/2015/aug/10/us-no-fly-list-predictive-assessments.

Adam, Barbara. 1994. *Time and Social Theory*. Cambridge: Polity Press.

———. 2004. *Time*. Cambridge: Polity Press.

Adey, Peter, Ben Anderson, and Stephen Graham. 2015. "Introduction: Governing Emergencies: Beyond Exceptionality." *Theory, Culture & Society* 32 (2): 3–17.

Agamben Giorgio. 1998. *Homo Sacer: Sovereign Power and Bare Life*. Stanford: Stanford University Press.

———. 2000. *Means without End: Notes on Politics*. Minneapolis: University of Minnesota Press.

———. 2002. "Security and Terror." *Theory and Event* 5 (4).

———. 2005. *State of Exception*. Chicago: University of Chicago Press.

Agathangelou, Anna, and Kyle Killian, eds. 2016. *Time, Temporality, and Violence in International Relations: De-fatalizing the Present, Forging Radical Alternatives*. New York: Routledge.

American Bar Association. 2006. *Task Force on Presidential Signing Statements and the Separation of Powers Doctrine Report*. Chicago: American Bar Association.

Amoore, Louise. 2008. "Consulting, Culture, and the Camp: On the Economies of Exception." In *Risk and the War on Terror*, edited by Louise Amoore and Marieke de Goede, 112–29. London: Routledge.

Amoore, Louise. 2014a. *Politics of Possibility: Risk and Security Beyond Probability*. Durham, NC: Duke University Press.

———. 2014b. "Security and the Incalculable." *Security Dialogue* 45 (5): 423–39.

Amoore, Louise, and Marieke de Goede, eds. 2008. *Risk and the War on Terror*. London: Routledge.

Ancona, Deborah, Paul S. Goodman, Barbara S. Lawrence, and Michael L. Tushman. 2001. "Time: A New Research Lens." *Academy of Management Review* 26 (4): 645–63.

Anderson, Ben. 2007. "Hope for Nanotechnology: Anticipatory Knowledge and the Governance of Affect." *Area* 39 (2): 156–65.

———. 2010a. "Pre-emption, Precaution, Preparedness: Anticipatory Action and Future Geographies." *Progress in Human Geography* 34 (6): 777–98.

———. 2010b. "Security and the Future: Anticipating the Event of Terror." *Geoforum* 41: 227–35.

Aradau, Claudia. 2015. "'Crowded Places are Everywhere We Go': Crowds, Emergency, Politics." *Theory, Culture & Society* 32 (2): 155–75.

Aradau, Claudia, Luis Lobo-Guerrero, and Rens van Munster. 2008. "Security, Technologies of Risk, and the Political: Guest Editors' Introduction." *Security Dialogue* 39 (2–3): 147–54.

Aradau, Claudia, and Rens van Munster. 2007. "Governing Terrorism Through Risk: Taking Precautions, (Un)Knowing the Future." *European Journal of International Relations* 13 (1): 89–115.

———. 2008. "Taming the Future: The *Dispositif* of Risk in the War on Terror." In *Risk and the War on Terror*, edited by Louise Amoore and Marieke de Goede, 23–40. London: Routledge.

———. 2009. "Exceptionalism and the 'War on Terror': Criminology Meets International Relations." *British Journal of Criminology* 49: 686–701.

———. 2011. *Politics of Catastrophe: Genealogies of the Unknown*. London: Routledge.

Arend, Anthony Clark. 2003. "International Law and the Preemptive Use of Military Force." *Washington Quarterly* 22 (2): 89–103.

Arfi, Badredine. 2010a. "'Euro-Islam': Going Beyond the Aporiatic Politics of Othering." *International Political Sociology* 4 (3): 236–52.

———. 2010b. "Rethinking International Constitutional Order: The Auto-Immune Politics of Binding Without Binding." *Millennium* 39 (2): 299–321.

Atwood, Margaret. 2008. *Payback: Debt and the Shadow Side of Wealth*. Toronto: House of Anansi Press.

Augustine of Hippo. 1968. "Some Questions about Time." In *The Philosophy of Time*, edited by R. M. Gale. London: Macmillan.

Bartelson, Jens. 1995. *A Genealogy of Sovereignty*. Cambridge: Cambridge University Press.

Bartosiewicz, Petra. 2012. "Deploying Informants, the FBI Stings Muslims." *Nation*, July 2–9. http://www.thenation.com/article/deploying-informants-fbi-stings-muslims/.

Bastian, Michelle. 2012. "Fatally Confused: Telling the Time in the Midst of Ecological Crises." *Environmental Philosophy* 9 (1): 23–48.

Bauman, Zygmunt. 2000. *Liquid Modernity*. Cambridge: Polity Press.

Beck, Ulrich. 1992. *Risk Society: Towards a New Modernity*. New York: Sage Publications.

———. 1999. *World Risk Society*. Cambridge: Polity Press.

———. 2002. "The Terrorist Threat: World Risk Society Revisited." *Theory, Culture & Society* 19 (4): 39–55.

——. 2005. "World Risk Society and the Changing Foundations of Transnational Politics." In *Complex Sovereignty: Reconstituting Political Authority in the Twenty-First Century*, edited by E. Grande and L. Pauly, 22–47. Toronto: University of Toronto Press.

———. 2008. *World at Risk*. Cambridge: Polity Press.

Becker, Jo, and Scott Shane. 2012. "A Measure of Change: Secret 'Kill List' Proves a Test of Obama's Principles and Will." *New York Times*, May 29. http://www.nytimes.com/2012/05/29/world/obamas-leadership-in-war-on-al-qaeda.html?_r=2&.

Bender, John, and David E. Wellbery, eds. 1991. *Chronotypes: The Construction of Time*. Stanford: Stanford University Press.

Benhabib, Seyla. 1994. "Deliberative Rationality and Models of Democratic Legitimacy." *Constellations* 1 (1): 26–52.

Bernstein, Peter. 1998. *Against the Gods: The Remarkable Story of Risk*. New York: John Wiley & Sons.

Bessant, Judith, Richard Hill, and Rob Watts. 2003. "*Discovering*" *Risk: Social Research and Policy Making*. New York: Peter Lang.

Best, Jacqueline. 2008. "Ambiguity, Uncertainty, and Risk: Rethinking Indeterminacy." *International Political Sociology* 2 (4): 355–74.

Betts, Richard K. 2003. "Striking First: A History of Thankfully Lost Opportunities." *Ethics & International Affairs* 17 (1): 17–24.

Blank, Laurie. 2015. "Targeted Killing." In *The Ashgate Research Companion to Military Ethics*, edited by J. Turner Johnson and E. D. Patterson, 227–43. Burlington, VT: Ashgate.

Bonelli, Laurent, and Francesco Ragazzi. 2014. "Low-Tech Security: Files, Notes, and Memos as Technologies of Anticipation." *Security Dialogue* 45 (5): 476–93.

Borradori, Giovanna. 2003. "Introduction: Terrorism and the Legacy of the Enlightenment." In *Philosophy in a Time of Terror: Dialogues with Jürgen Habermas and Jacques Derrida*, edited by G. Borradori, 1–22. Chicago: University of Chicago Press.

Bridge, Robert. 2012. "Putin: Russia Must be Unyielding Against Terrorists." *RT Online*, October 16. http://rt.com/politics/putin-russia-terrorists-olympics-sochi-563/.

Bull, Hedley. 1977. *The Anarchical Society*. New York: Columbia University Press.

Bush, George W. 2006. "Statement on H. R. 1815, 'National Defense Authorization Act for Fiscal Year 2006.'" Washington, DC: White House.

Butler, Judith. 2006. *Precarious Life: The Powers of Mourning and Violence*. London: Verso.

Campbell, David. 1998. *Writing Security*. Minneapolis: University of Minnesota Press.

Canadian International Council. 2012. "A Drone Field Guide." opencanada.org, May 31. http://opencanada.org/features/the-think-tank/graphic/a-drone-field-guide/.

Canadian Security and Intelligence Service. 2011. *2010–2011 Public Report*. Ottawa: Government of Canada. http://www.csis-scrs.gc.ca/pblctns/nnlrprt/2010-2011/rprt2010-2011-eng_final.asp.

CBS. 2012. "Defense Secretary Panetta." *60 Minutes*, January 29. http://www.cbsnews.com/8301-18560_162-57367997/the-defense-secretary-leon-panetta/.

Cheah, Pheng, and Suzanne Guerlac. 2009. "Introduction: Derrida and the Time of the Political." In *Derrida and the Time of the Political*, edited by P. Cheah and S. Guerlac, 1–37. Durham, NC: Duke University Press.

Clough, David, and Brian Stiltner. 2007. "On the Importance of the Drawn Sword: Christian Thinking about Preemptive War—and its Modern Outworking." *Journal of the Society of Christian Ethics* 27 (2): 253–71.

Coaffee, Jon. 2009. *Terrorism, Risk, and the Global City*. London: Ashgate.

Cohen, E. 2004. "My Self as an Other: On Autoimmunity and 'Other' Paradoxes." *Medical Humanities* 30 (1): 7–11.

Collier, Stephen J., and Andrew Lakoff. 2015. "Vital Systems Security: Reflexive Biopolitics and the Government of Emergency." *Theory, Culture & Society* 32 (2): 19–51.

Connolly, William E. 2002. *Neuropolitics: Thinking, Culture, Speed*. Minneapolis: University of Minnesota Press.

Cooper, Melinda. 2006. "Pre-empting Emergence: The Biological Turn in the War on Terror." *Theory, Culture & Society* 23 (4): 113–35.

———. 2015. "The Theology of Emergency: Welfare Reform, US Foreign Aid, and the Faith-Based Initiative." *Theory, Culture & Society* 32 (2): 53–77.

Crang, Mike. 2007. "Speed = Distance/Time: Chronotopographies of Action." In *24/7: Time and Temporality in the Network Society*, edited by R. Hassan and R. Purser, 62–88. Stanford: Stanford University Press.

Daase, Christopher, and Oliver Kessler. 2007. "Knowns and Unknowns in the 'War on Terror': Uncertainty and the Political Construction of Danger." *Security Dialogue* 38 (4): 411–34.

de Goede, Marieke. 2008. "The Politics of Pre-emption and the War on Terror in Europe." *European Journal of International Relations* 14 (1): 161–85.

———. 2011. "Blacklisting and the Ban: Contesting Targeted Sanctions in Europe." *Security Dialogue* 42 (6): 499–515.

———. 2012. *Speculative Security: The Politics of Pursuing Terrorist Monies*. Minneapolis: University of Minnesota Press.

de Goede, Marieke, and Samuel Randalls. 2009. "Precaution, Pre-emption: Arts and Technologies of the Actionable Future." *Environment and Planning D: Society and Space* 27: 859–78.

de Goede, Marieke, Stephanie Simon, and Marijn Hoijtnik. 2014. "Performing Preemption." *Security Dialogue* 45 (5): 411–22.

Der Derian, James. 2008. "Becoming Connolly: Critique, Crossing Over, and Concepts." In *The New Pluralism: William Connolly and the Contemporary Global Condition*, edited by D. Campbell and M. Schoolman, 289–304. Durham, NC: Duke University Press.

———. 2009. "The Value of Security: Hobbes, Marx, Nietzsche, Baudrillard." In *Critical Practices in International Theory: Selected Essays*, 149–66. London: Routledge.

Derrida, Jacques. 1990. "Force of Law: The Mystical Foundation of Authority." *Cardozo Law Review* 11: 920–1045.

———. 2005. *Rogues: Two Essays on Reason*. Stanford: Stanford University Press.

———. 2003. "Autoimmunity: Real and Symbolic Suicides." In *Philosophy in a Time of Terror: Dialogues with Jürgen Habermas and Jacques Derrida*, edited by G. Borradori, 85–136. Chicago: University of Chicago Press.

Dershowitz, Alan. 2006. *Pre-emption: A Knife that Cuts Both Ways*. New York: Norton.

Dillon, Michael. 2008. "Underwriting Security." *Security Dialogue* 39 (2–3): 309–32.

Diprose, Rosalyn, Niamh Stephenson, Catherine Mills, Kane Race, and Gay Hawkins. 2008. "Governing the Future: The Paradigm of Prudence in Political Technologies of Risk Management." *Security Dialogue* 39 (2–3): 267–88.

Doty, Roxanne Lynn. 2007. "States of Exception on the US-Mexico Border: Security, 'Decisions,' and Civilian Border Patrols." *International Political Sociology* 1 (2): 113–37.

———. 2009. *The Law into Their Own Hands: Immigration and the Politics of Exceptionalism*. Tucson: University of Arizona Press.

Doyle, Michael W. 2006. *Striking First: Pre-emption and Prevention in International Conflict*. Princeton, NJ: Princeton University Press.

Dozier, Kimberly. 2012. "Who Will Drones Target? Who in the US Will Decide?" *Salon*, May 21. http://www.salon.com/2012/05/21/who_will_drones_target_who_in_the_us_will_decide/.

Edkins, Jenny. 2000. "Sovereign Power, Zones of Indistinction, and the Camp." *Alternatives* 25: 3–25.

———. 2002. "Forget Trauma?" *International Relations* 16 (2): 243–56.

Edkins, Jenny, and Nick Vaughan-Williams. 2009. "Introduction." In *Critical Theorists and International Relations*, edited by J. Edkins and N. Vaughan-Williams, 1–6. New York: Routledge.

Ehrenberg, John, J. Patrice McSherry, José Ramón Sánchez, and Caroleen Marji Sayej, eds. 2010. *The Iraq Papers*. Oxford: Oxford University Press.

Elias, Norbert. 1992. *Time: An Essay*. Oxford: Oxford University Press.

Elmer, Greg, and Andy Opel. 2006. "Surviving the Inevitable Future: Pre-emption in an Age of Faulty Intelligence." *Cultural Studies* 20 (4–5): 477–92.

———. 2008. *Pre-empting Dissent: The Politics of an Inevitable Future*. Winnipeg: Arbeiter Ring Publishers.

Elshtain, Jean Bethke. 2013. "Prevention, Preemption, and Other Conundrums." In *The Ethics of Preventive War*, edited by D. K. Chatterjee, 15–26. Cambridge: Cambridge University Press.

Ericson, Richard. 2007. *Crime in an Insecure World*. Cambridge: Polity Press.

Ericson, Richard V. 2008. "The State of Pre-emption: Managing Terrorism Risk Through Counter Law." In *Risk and the War on Terror*, edited by L. Amoore and M. de Goede, 57–76. London: Routledge.

Ericson, Richard V., and Kevin D. Haggerty. 2002. "The Policing of Risk." In *Embracing Risk: The Changing Culture of Insurance and Responsibility*, edited by T. Baker and J. Simon, 238–72. Chicago: University of Chicago Press.

European Commission. 2000. *Communication from the Commission on the Precautionary Principle*. Brussels: Commission for the European Communities.

European Union. 2013. "European Commission Enlargement Policy—Conditions for Membership." May 7. http://ec.europa.eu/enlargement/policy/conditions-membership/.

Evans, Gareth. 2004. "When Is It Right to Fight?" *Survival* 46 (3): 59–82.

Ewald, François. 1991. "Insurance and Risk." In *The Foucault Effect: Studies in Governmentality*, edited by G. Burchill, C. Gordon, and P. Miller, 197–210. Chicago: University of Chicago Press.

———. 1993. "Two Infinities of Risk." In *The Politics of Everyday Fear*, edited by B. Massumi, 221–28. Minneapolis: University of Minnesota Press.

———. 2002. "The Return of Descartes' Malicious Demon: An Outline of a Philosophy of Precaution." In *Embracing Risk: The Changing Culture of Insurance and Responsibility*, edited by T. Baker and J. Simon, 273–301. Chicago: University of Chicago Press.

Fabian, Johannes. 2002. *Time and the Other: How Anthropology Makes Its Object*. New York: Columbia University Press.

Falk, Barbara J. 2010. "From Berlin to Baghdad: Learning the 'Wrong' Lessons from the Collapse of Communism." In *The Global 1989*, edited by G. Lawson, C. Armbruster, and M. Cox, 243–70. Cambridge: Cambridge University Press.

Federal Bureau of Investigation. 2015. "3 Brooklyn Residents Charged with Attempt and Conspiracy to Provide Material Support to ISIL." Press Release: New York Field Office, February 25. https://www.fbi.gov/newyork/press-releases/2015/three-brooklyn-residents-charged-with-attempt-and-conspiracy-to-provide-material-support-to-isil.

Fierke, Karin. 2007. *Critical Approaches to International Security*. Cambridge: Polity Press.

Fisher, Philip. 2002. *The Vehement Passions*. Princeton, NJ: Princeton University Press.

Fletcher, George P., and Jens D. Ohlin. 2008. *Defending Humanity: When Force is Justified and Why*. New York: Oxford University Press.

Foucault, Michel. 1980. *Power/Knowledge*. New York: Pantheon.

———. 1991. "Governmentality." In *The Foucault Effect: Studies in Governmentality*, edited by G. Burchill, C. Gordon, and P. Miller, 87–104. Chicago: University of Chicago Press.

Freeze, Colin, and Greg McArthur. 2011. "Man Jailed in New Kind of Terrorist Case for Canada." *Globe and Mail*, March 30. http://www.theglobeandmail.com/news/national/toronto/rcmp-arrest-terror-suspect-about-to-board-plane-in-toronto/article1962875/.

Furedi, Frank. 2009. "Precautionary Culture and the Rise of Possibilistic Risk Assessment." *Erasmus Law Review* 2 (2): 197–220.

Gale, Richard M., ed. 1968. *The Philosophy of Time*. London: Macmillan.

Gardner, Dan. 2009. *Risk: The Science and Politics of Fear*. Toronto: McClelland and Stewart.

Garland, David. 2003. "The Rise of Risk." In *Risk and Morality*, edited by R. Ericson and A. Doyle, 48–86. Toronto: University of Toronto Press.

Gasché, Rodolphe. 2004. "'In the Name of Reason': The Deconstruction of Sovereignty." *Research in Phenomenology* 34: 289–303.

Gelev, Filip. 2011. "Checks and Balances of Risk Management: Precautionary Logic and the Judiciary." *Review of International Studies* 37: 2237–52.

Giddens, Anthony. 1984. *The Constitution of Society: Introduction to the Theory of Structuration*. Berkeley: University of California Press.

———. 1991. *Modernity and Self-Identity: Self and Society in the Late Modern Age*. Cambridge: Polity Press.

Government of Canada. 2011. *Building Resilience Against Terrorism: Canada's Counter-Terrorism Strategy*. Ottawa: Government of Canada.

Grayson, Kyle. 2012. "Six Theses on Targeted Killing." *Politics* 32 (2): 120–28.

Greenwald, Glenn. 2012. "Attorney General Holder Defends Execution Without Charges." *Salon*, March 6. http://www.salon.com/2012/03/06/attorney_general_holder_defends_execution_without_charges.

Greenwald, Glenn, and Andrew Fishman. 2015. "Latest FBI Claim of Disrupted Terror Plot Deserves Much Scrutiny and Skepticism." *Intercept*, January 16. https://firstlook

.org/theintercept/2015/01/16/latest-fbi-boast-disrupting-terror-u-s-plot-deserves-scrutiny-skepticism/.

Grosz, Elizabeth, ed. 1999. *Beginnings: Explorations in Time, Memory, and Futures*. Ithaca: Cornell University Press.

Grotius, Hugo. 2001 [1625]. *On the Law of War and Peace*. Translated by A. C. Campbell. Kitchener, Ontario: Batoche Books.

———. 2012 [1625]. *On the Law of War and Peace*. Edited by Stephen C. Neff. Cambridge: Cambridge University Press.

Guertner, Gary. 2007. "European Views of Preemption in US National Security Strategy." *Parameters* 37 (2): 31–44.

Gunnell, John. 1987. *Political Philosophy and Time: Plato and the Origins of Political Vision*. Chicago: University of Chicago Press.

Hacking, Ian. 1990. *The Taming of Chance*. Cambridge: Cambridge University Press.

Hahn, Robert, and Cass Sunstein. 2005. "The Precautionary Principles as a Basis for Decision-Making." *Economist's Voice* 2 (2): 1–9.

Harvey, David. 1991. *The Condition of Postmodernity: An Enquiry into the Origins of Cultural Change*. Oxford: Blackwell.

Hassan, Robert, and Ronald Purser, eds. 2007. *24/7: Time and Temporality in the Network Society*. Stanford: Stanford University Press.

Hobbes, Thomas. 2011 [1651]. *Leviathan*. Edited by A. P. Martinich and Brian Battiste. Peterborough, Ontario: Broadview Editions.

Hoffman, John. 1998. *Sovereignty*. Minneapolis: University of Minnesota Press.

Honig, Bonnie. 2008. "The Time of Rights: Emergency Thoughts in an Emergency Setting." In *The New Pluralism: William Connolly and the Contemporary Global Condition*, edited by D. Campbell and M. Schoolman, 85–120. Durham, NC: Duke University Press.

Human Rights Watch. 2014. *Illusion of Justice: Human Rights Abuses in US Terrorism Prosecutions*. New York: Columbia Law School Human Rights Institute.

Humphreys, Adrian. 2014. "Canadian 'Terror Tourist' Mohamed Hersi Gets 10 Years in Jail for Planning to Join Islamic Jihadist Group in Somalia." *National Post*, July 24. http://news.nationalpost.com/news/canada/canadian-terror-tourist-mohamed-hersi-gets-10-years-in-jail-for-planning-to-join-islamic-jihadist-group-in-somalia.

Hussain, Murtaza. 2015. "Confidential Informant Played Key Role in FBI Foiling its Own Terror Plot." *Intercept*, February 25. https://firstlook.org/theintercept/2015/02/25/isis-material-support-plot-involved-confidential-informant/.

Hutchings, Kimberly. 2008. *Time and World Politics: Thinking the Present*. Manchester: Manchester University Press.

Huysmans, Jef. 2004. "Minding Exceptions: The Politics of Insecurity and Liberal Democracy." *Contemporary Political Theory* 3: 321–41.

———. 2006. "International Politics of Exception: Competing Visions of International Political Order Between Law and Politics." *Alternatives* 31: 135–65.

———. 2008. "The Jargon of Exception—On Schmitt, Agamben, and the Absence of Political Society." *International Political Sociology* 2 (3): 165–83.

Isin, Engin. 2004. "The Neurotic Citizen." *Citizenship Studies* 8 (3): 217–35.

Isin, Engin, and Kim Rygiel. 2007. "Abject Spaces: Frontiers, Zones, Camps." In *The Logics of Biopower and the War on Terror: Living, Dying, Surviving*, edited by E. Dauphinée and C. Masters, 181–203. Houndmills, UK: Palgrave Macmillan.

Jackson, Robert. 2007. *Sovereignty*. Cambridge: Polity Press.

Jarvis, Darryl. 2007. "Risk, Globalization, and the State: A Critical Appraisal of Ulrich Beck and the World Risk Society Thesis." *Global Society* 21 (1): 23–46.

Jarvis, Lee. 2009. *Times of Terror: Discourse, Temporality, and the War on Terror*. London: Palgrave Macmillan.

Johns, Fleur. 2005. "Guantánamo Bay and the Annihilation of the Exception." *European Journal of International Law* 16 (4): 613–35.

Kalhan, Anil, Gerald Conroy, Mamta Kaushal, Sam Scott Miller, and Jed Rakoff. 2007. "Colonial Continuities: Human Rights, Terrorism, and Security Laws in India." *Columbia Journal of Asian Law* 20 (1): 93–234.

Kegley, Charles W., and Gregory A. Raymond. 2004. "Global Terrorism and Military Preemption: Policy Problems and Normative Perils." *International Politics* 41: 37–49.

Kessler, Oliver. 2011. "The Same as it Never Was? Uncertainty and the Changing Contours of International Law." *Review of International Studies* 37: 2163–82.

Kessler, Oliver, and Christopher Daase. 2008. "From Insecurity to Uncertainty: Risk and the Paradox of Security Politics." *Alternatives* 33: 211–32.

Kessler, Oliver, and Wouter Werner. 2008. "Extrajudicial Killing as Risk Management." *Security Dialogue* 39 (2–3): 289–308.

Kindred, Hugh M., and Phillip Sanders, eds. 2006. *Public International Law Chiefly as Applied and Interpreted in Canada*, 7th ed. Toronto: Emond Montgomery.

Knight, Frank. 1946. *Risk, Uncertainty and Profit*. Boston: Houghton Mifflin.

Koring, Paul. 2011. "Memo Equips Obama with 'Licence to Kill.'" *Globe and Mail*, October 9. http://www.theglobeandmail.com/news/world/americas/memo-equips-obama-with-licence-to-kill/article2196145/.

Koselleck, Rinehart. 1985. *Futures Past: On the Semantics of Historical Time*. New York: Columbia University Press.

Krishnan, Jayanth. 2004. "India's Patriot Act: POTA and the Impact on Civil Liberties in the World's Largest Democracy." *Law and Inequality* 22: 265–300.

Kurtulus, Ersun. 2005. *State Sovereignty: Concept, Phenomenon, and Ramifications*. New York: Palgrave Macmillan.

———. 2007. "The Notion of a 'Pre-Emptive War': The Six Day War Revisited." *Middle East Journal* 61 (2): 220–38.

Lazar, Nomi Claire. 2006. "Must Exceptions Prove the Rule? An Angle on Emergency Government in the History of Political Thought." *Politics & Society* 34 (2): 245–75.

———. 2009. *States of Emergency in Liberal Democracy*. Cambridge: Cambridge University Press.

Leander, Anna. 2011. "Risk and the Fabrication of Apolitical, Unaccountable Security Markets: The Case of the CIA 'Killing Program.'" *Review of International Studies* 37: 2253–68.

Leccardi, Carmen. 2007. "New Temporal Perspectives in the 'High-Speed Society.'" In *24/7: Time and Temporality in the Network Society*, edited by R. Hassan and R. Purser, 25–36. Stanford: Stanford University Press.

Leese, Matthias. 2014. "The New Profiling: Algorithms, Black Boxes, and the Failure of Anti-discriminatory Safeguards in the European Union." *Security Dialogue* 45 (5): 494–511.

Luhmann, Niklas. 1982. *The Differentiation of Society*. New York: Columbia University Press.

Lundborg, Tom. 2012. *Politics of the Event: Time, Movement, Becoming*. New York: Routledge.
Lupton, Deborah. 1999. *Risk*. London: Routledge.
Machiavelli, Niccolo. 1995. *The Prince*. Translated by David Wootton. Indianapolis, IN: Hackett.
MacNaghten, Phil. 2005. "Environment and Risk." In *Beyond the Risk Society: Critical Reflections on Risk and Human Security*, edited by G. Mythen and S. Walklate, 132–43. London: Open University Press.
Manning, Erin. 2004. "Time for Politics." In *Sovereign Lives: Power in Global Politics*, edited by J. Edkins, V. Pin-Fat, and M. Shapiro, 61–78. London: Routledge.
Massey, Doreen. 1994. *Space, Place, and Gender*. Minneapolis: University of Minnesota Press.
Massumi, Brian. 2005a. "The Future Birth of the Affective Fact." *Conference Proceedings: Genealogies of Biopolitics*, 1–12.
———. 2005b. "Fear (The Spectrum Said)." *Positions* 13 (1): 31–48.
———. 2007. "Potential Politics and the Primacy of Pre-emption." *Theory & Event* 10 (2).
———. 2010. "The Future Birth of the Affective Fact: The Political Ontology of Threat." In *The Affect Theory Reader*, edited by M. Gregg and G. Seigworth, 52–70. Durham, NC: Duke University Press.
Mate, Manoj, and Adnan Naseemullah. 2010. "State Security and Elite Capture: The Implementation of Anti-Terrorist Legislation in India." *Journal of Human Rights* 9: 262–78.
McClure, Roger. 2005. *The Philosophy of Time: Time before Times*. New York: Routledge.
McCumber, John. 2011. *Time and Philosophy: A History of Continental Thought*. Montreal: McGill-Queen's University Press.
Merriam-Webster. 2006. "Word of the Year Archive: 2006 Word of the Year." http://www.merriam-webster.com/word-of-the-year/2006-word-of-the-year.htm.
Miller, J. Hillis. 2009. *For Derrida*. New York: Fordham University Press.
Mitchell, W. J. T. 2005. "Picturing Terror: Derrida's Autoimmunity." *Cardozo Law Review* 27 (2): 913–92.
Muller, Benjamin. 2010. *Security, Risk, and the Biometric State: Governing Borders and Bodies*. London: Routledge.
Mutimer, David. 2007. "Sovereign Contradictions: Maher Arar and the Indefinite Future." In *The Logics of Biopower and the War on Terror: Living, Dying Surviving*, edited by E. Dauphinée and C. Masters, 159–79. Houndmills, UK: Palgrave.
Mythen, Gabe. 2005. *Ulrich Beck: A Critical Introduction to the Risk Society*. London: Pluto Press.
Naas, Michael. 2006. "'One Nation . . . Indivisible': Jacques Derrida on the Autoimmunity of Democracy and the Sovereignty of God." *Research in Phenomenology* 36: 15–44.
Neal, Andrew. 2009. *Exceptionalism and the Politics of Counter-Terrorism: Liberty, Security, and the War on Terror*. New York: Routledge.
Nelson, Scott. 2010. *Sovereignty and the Limits of the Liberal Imagination*. New York: Routledge.
Nichols, Thomas M. 2008. *Eve of Destruction: The Coming Age of Preventive War*. Philadelphia: University of Pennsylvania Press.

Nietzsche, Friedrich. 1989. *On the Genealogy of Morals*. Translated by W. Kaufmann. New York: Vintage Books.

Nyers, Peter. 2006. "The Accidental Citizen: Acts of Sovereignty and (un)Making Citizenship." *Economy and Society* 35 (1): 22–41.

O'Hanlon, Michael, Susan Rice, and James Steinberg. 2002. "The New National Security Strategy and Pre-emption." *Brookings Institution Policy Brief Series*, no. 113. http://www.brookings.edu/research/papers/2002/12/terrorism-ohanlon.

O'Malley, Pat. 2005. "Criminology and Risk." In *Beyond the Risk Society: Critical Reflections on Risk and Human Security*, edited by G. Mythen and S. Walklate, 43–59. London: Open University Press.

Onuf, Nicholas. 1991. "Sovereignty: Outline of a Conceptual History." *Alternatives* 16 (4): 425–46.

Opitz, Sven, and Ute Tellmann. 2015. "Future Emergencies: Temporal Politics in Law and Economy." *Theory, Culture & Society* 32 (2): 107–29.

Pocock, J. G. A. 2003. *The Machiavellian Moment: Florentine Political Thought and the Atlantic Republican Tradition*. Princeton, NJ: Princeton University Press.

Porter, Tony. 2009. "Risk Models and Transnational Governance in the Global Financial Crisis: The Cases of Basel II and Credit Rating Agencies." In *Global Finance in Crisis: The Politics of International Regulatory Change*, edited by E. Helleiner, S. Pagliari, and H. Zimmermann, 56–73. London: Routledge.

Power, Michael. 2004. *The Risk Management of Everything: Rethinking the Politics of Uncertainty*. London: Demos.

Prigogine, Ilya. 1997. *The End of Certainty*. New York: Free Press.

Prokhovnik, Raia. 2007. *Sovereignties: Contemporary Theory and Practice*. New York: Palgrave.

Prozorov, Sergei. 2005. "X/Xs: Toward a General Theory of the Exception." *Alternatives* 30: 81–111.

Pufendorf, Samuel. 1991 [1673]. *On the Duty of Man and Citizen According to Natural Law*. Edited by James Tully. Cambridge: Cambridge University Press.

Ricoeur, Paul. 1990. *Time and Narrative*. Chicago: University of Chicago Press.

Rosa, Hartmut. 2009. "Social Acceleration: Ethical and Political Consequences of a Desynchronized High-Speed Society." In *High-Speed Society: Social Acceleration, Power, and Modernity*, edited by H. Rosa and W. Scheuerman. University Park: Pennsylvania State University Press.

———. 2013. *Social Acceleration: A New Theory of Modernity*. New York: Columbia University Press.

Rosa, Hartmut, and William Scheuerman, eds. 2009. *High-Speed Society: Social Acceleration, Power, and Modernity*. University Park: Pennsylvania State University Press.

Rose, Nikolas. 2001. "The Politics of Life Itself." *Theory Culture & Society* 28 (6): 1–30.

Rose, Nikolas, and Peter Miller. 1992. "Political Power beyond the State: Problematics of Government." *British Journal of Sociology* 43 (2): 173–205.

Runciman, David. 2004. "The Precautionary Principle." *London Review of Books* 26 (7), April 1. http://www.lrb.co.uk/v26/n07/david-runciman/the-precautionary-principle.

Salter, Mark B. 2008a. "Risk and Imagination in the War on Terror." In *Risk and the War on Terror*, edited by M. de Goede and L. Amoore, 233–47. London: Routledge.

———. 2008b. "When the Exception Becomes the Rule: Borders, Sovereignty, and Citizenship." *Citizenship Studies* 12 (4): 365–80.

Santora, Marc, and Stephanie Clifford. 2015. "3 Brooklyn Men Accused of Plot to Aid ISIS' Fight." *New York Times*, February 25. http://www.nytimes.com/2015/02/26/nyregion/3-men-in-brooklyn-charged-supporting-isis.html?_r = 0.

Sasikumar, Karthika. 2010. "State Agency in the Time of the Global War on Terror: India and the Counter-Terrorism Regime." *Review of International Studies* 36: 615–38.

Schmitt, Carl. 2005. *Political Theology: Four Chapters on the Concept of Sovereignty*. Translated by G. Schwab. Chicago: University of Chicago Press.

Schwab, George. 2005. "Introduction." In *Political Theology: Four Chapters on the Concept of Sovereignty*, by Carl Schmitt, translated by G. Schwab, xxxvii–lii. Chicago: University of Chicago Press.

Slaughter, Anne-Marie. 2004. "Disaggregated Sovereignty: Toward the Public Accountability of Global Government Networks." *Government and Opposition* 39 (2): 159–90.

Spruyt, Hendrik. 1996. *The Sovereign State and its Competitors*. Princeton, NJ: Princeton University Press.

Stankiewicz, W. J. 1969. "In Defense of Sovereignty: A Critique and an Interpretation." In *In Defense of Sovereignty*, edited by W. J. Stankiewicz, 3–38. New York: Oxford University Press.

Stern, Jessica, and Jonathan Wiener. 2006. "Precaution Against Terrorism." *Journal of Risk Research* 9 (4): 393–447.

Stockdale, L. P. D. 2013 "Imagined Futures and Exceptional Presents: A Conceptual Critique of 'Pre-emptive Security.'" *Global Change, Peace & Security* 25 (2): 141–57.

Strong, Tracy B. 2005. "Foreword: The Sovereign and the Exception: Carl Schmitt, Politics, Theology, and Leadership." In *Political Theology: Four Chapters on the Concept of Sovereignty*, by Carl Schmitt, translated by G. Schwab, vii–xxxv. Chicago: University of Chicago Press.

Sullivan, Gavin, and Ben Hayes. 2010. *Blacklisted: Targeted Sanctions, Pre-emptive Security, and Human Rights*. Berlin: European Centre for Constitutional and Human Rights.

Sunstein, Cass. 2007. *Worst-Case Scenarios*. Cambridge, MA: Harvard University Press.

Svendsen, Adam. 2010. "Re-fashioning Risk: Comparing UK, US, and Canadian Security and Intelligence Efforts Against Terrorism." *Defence Studies* 10 (3): 307–35.

Taleb, Nassim Nicholas. 2010. *The Black Swan: The Impact of the Highly Improbable*, 2nd ed. New York: Random House.

Taylor, Peter. 2006. "The Terrorist Who Wasn't." *Guardian*, March 8. http://www.guardian.co.uk/uk/2006/mar/08/menezes.july7.

Taylor-Gooby, Peter, and Jens Zinn. 2006. "Risk as an Interdisciplinary Research Area." In *Risk in Social Science*, edited by P. Taylor-Gooby and J. O. Zinn, 20–53. Oxford: Oxford University Press.

Teotonio, Isabel. 2011. "U of T Graduate's Arrest on Terror Charges Alarms Toronto Somalis." *Toronto Star*, March 31. http://www.thestar.com/news/crime/article/965595--toronto-man-arrested-on-terrorism-related-charges.

Turetzky, Philip. 1998. *Time*. New York: Routledge.

United Nations. 1992. *Declaration on Environment and Development*. New York: United Nations General Assembly.

———. 2004. *A More Secure World: Our Shared Responsibility—Report of the High-Level Panel on Threats, Challenges and Change*. New York: United Nations General Assembly, September 28.

———. 2013. *Disaster Risk Reduction in the United Nations*. New York: United Nations Office for Disaster Risk Reduction.

United Nations Security Council. 2001. *Resolution 1373*. New York: United Nations, September 28. http://www.un.org/en/sc/ctc/specialmeetings/2012/docs/United%20Nations%20Security%20Council%20Resolution%201373%20(2001).pdf.

United States v. Cromitie. 781 F. Supp. 2d 211 (S.D.N.Y. 2011).

United States Department of Homeland Security. 2011. "Risk Management Fundamentals: Homeland Security Risk Management Doctrine." Washington, DC: US Department of Homeland Security.

United States Department of Justice. 2006. "Legal Authorities Supporting the Activities of the National Security Agency Described by the President." *Department of Justice White Paper*. Washington, DC: US Department of Justice. http://www.justice.gov/opa/whitepaperonnsalegalauthorities.pdf.

———. 2012. "Attorney General Eric Holder Speaks at Northwestern University School of Law." Washington, DC: US Department of Justice. March 5. http://www.justice.gov/iso/opa/ag/speeches/2012/ag-speech-1203051.html.

———. 2013. "Lawfulness of a Lethal Operation Against a US Citizen Who Is a Senior Operational Leader of Al-Qa'ida or an Associated Force." *Department of Justice White Paper*. Washington, DC: US Department of Justice.

Urry, John. 2003. *Global Complexity*. Cambridge: Polity Press.

van Munster, Rens. 2004. "The War on Terrorism: When the Exception Becomes the Rule." *International Journal for the Semiotics of Law* 17: 141–53.

Vaughan-Williams, Nick. 2007. "The Shooting of Jean Charles de Menezes: New Border Politics?" *Alternatives* 32 (2): 177–95.

Virilio, Paul. 2007. *The Original Accident*. Cambridge: Polity Press.

———. 2010. *The University of Disaster*. Cambridge: Polity Press.

Walker, R. B. J. 1991. "State Sovereignty and the Articulation of Political Space/Time." *Millennium* 20 (3): 445–61.

———. 1993. *Inside/Outside: International Relations as Political Theory*. Cambridge: Cambridge University Press.

———. 2004. "Sovereignties, Exceptions, Worlds." In *Sovereign Lives: Power in Global Politics*, edited by J. Edkins, V. Pin-Fat, and M. Shapiro, 239–49. London: Routledge.

———. 2006. "Lines of Insecurity: International, Imperial, Exceptional." *Security Dialogue* 37 (1): 65–82.

——. 2010. *After the Globe/Before the World*. London: Routledge.

Walzer, Michael. 1977. *Just and Unjust Wars: A Moral Argument with Historical Illustrations*. New York: Basic Books.

Weber, Cynthia. 1995. *Simulating Sovereignty: Intervention, the State, and Symbolic Exchange*. Cambridge: Cambridge University Press.

———. 2007. "Securitizing the Unconscious: The Bush Doctrine of Pre-emption and *Minority Report*." In *The Logics of Biopower and the War on Terror: Living, Dying Surviving*, edited by E. Dauphinée and C. Masters, 109–28. Houndmills, UK: Palgrave Macmillan.

Weber, Leanne. 2007. "Policing the Virtual Border: Punitive Pre-emption in Australian Offshore Migration Control." *Social Justice* 34 (2): 77–93.

Westphal, Steven D. 2003. "Counterterrorism: Policy of Preemptive Action." *United States Army War College Strategy Research Project*. Carlisle Barracks, PA: US Army War College.

White, Hayden. 1973. *Metahistory: The Historical Imagination in Nineteenth-Century Europe*. Baltimore: Johns Hopkins University Press.

White House. 2002. "National Security Strategy of the United States of America." Washington, DC: White House.

Wolin, Sheldon. 1997. "What Time Is It?" *Theory & Event* 1 (1).

———. 2004. *Politics and Vision*. Princeton, NJ: Princeton University Press.

Zehfuss, Maja. 2003. "Forget September 11." *Third World Quarterly* 23 (3): 513–38.

Index

About the Author

Liam P. D. Stockdale is a postdoctoral fellow jointly appointed to the Institute on Globalization and the Human Condition and the Institute for Innovation and Excellence in Teaching and Learning at McMaster University, Hamilton, Ontario, Canada. His research interests sit at the intersection of international relations and contemporary social and political theory, with an emphasis on the post-9/11 global politics of (in)security. He has also written and published on the international politics of sport and the temporalities of globalization.

www.ingramcontent.com/pod-product-compliance
Lightning Source LLC
LaVergne TN
LVHW101447200926
848483LV00001B/82
* 9 7 8 1 7 8 3 4 8 5 0 0 0 *